*Clarke War Memorial Fountain, University of Notre Dame, Notre Dame, Indiana*

Large blocks of Indiana limestone, 20 ft high and 5 ft$^2$, are set in "Stonehenge"-type fashion, capped with 10 ft slabs of the same stone in a fountain pool. The columns in the center are coarsely hewn, while the finish of the outer columns is sawed. The Indiana limestone, calcareous fossil fragments cemented with oolites, shows two color tones, the original bluish-gray color, and oxidized ochre, well seen on the blocks in the center. The blocks stand on end immersed in the shallow water pool; the bedding planes and cross-bedding appear vertical, well seen on the right-hand block. Continuous, wet/dry cycles with water introduced from four fountains positioned between the four pairs of blocks provide ample water to the stone. Secondary surface deposits of iron on the outer blocks and the white margins of gypsum are the result of interaction of the atmospheric sulfur with the limestone.

The pool basin is veneered with Italian black granite (gabbro) facing a solid concrete core. White seams on many joints are secondary crusts of white calcite leached from the concrete from water supplied by the pool behind under a constant head. The fountain started operation May 1986.

E.M. Winkler

# Stone in Architecture

## Properties, Durability

Third, Completely Revised and Extended Edition

With 219 Figures, Some in Color and 63 Tables

Springer-Verlag
Berlin Heidelberg New York
London Paris Tokyo
Hong Kong Barcelona
Budapest

Prof. Dr. Erhard M. Winkler
University of Notre Dame
Notre Dame, IN 46556-0767, USA

Third Edition
ISBN 3-540-57626-6 Springer-Verlag Berlin Heidelberg New York
ISBN 0-387-57626-6 Springer-Verlag New York Berlin Heidelberg

Second Edition
ISBN 3-211-81071-4 Springer-Verlag Wien New York

Library of Congress Cataloging-in-Publication Data. Winkler, Erhard M. Stone in architecture : properties, durability / Erhard M. Winkler.—3rd ed. p. cm. Rev. ed. of: Stone—properties, durability in man's environment. Includes bibliographical references and index. ISBN 0–387–57626–6 1. Stone. 2. Building stones. I. Winkler, Erhard M. Stone—properties, durability in man's environment. II. Title. TA426.W55 1994 624.1′832—dc20 94-20332

Typesetting: Best-set Typesetter Ltd., Hong Kong

SPIN: 10078217        32/3130/SPS – 5 4 3 2 1 0 – Printed on acid-free paper

To my wife Isolde

# Preface

The readers of the first two editions of Stone: Properties, Durability in Man's Environment, were mostly architects, restoration architects of buildings and monuments in natural stone, professionals who sought basic technical information for non-geologists. The increasing awareness of rapidly decaying monuments and their rescue from loss to future generations have urged this writer to update the 1973 and 1975 editions, now unavailable and out of print. Due to the 20-year-long interval, extensive updating was necessary to produce this new book.

The present edition concentrates on the natural material stone, as building stone, dimension stone, architectural stone, and decorative field stones. Recently, the use of stone for thin curtain walls on buildings has become fashionable. The thin slabs exposed to a new, unknown complexity of stresses, resulting in bowing of crystalline marble, has attracted much negative publicity. The costs of replacing white slabs of marble on entire buildings with its legal implications have led construction companies into bankruptcy. We blame many environmental problems on acid rain. Does acid rain really accelerate stone decay that much?

Stone preservation is being attempted with an ever-increasing number of chemicals applied by as many specialists to save crumbling stone. Chemists filled this need during a time of temporary job scarcity, while the general geologist missed this opportunity; he was too deeply involved in the search for fossil fuels and metals. A solid background in the broad field of general geology is required for a full understanding of the process of weathering in rural or urban macro- and microenvironments. For example, the bowing of thin panels of crystalline marble as claddings must be understood in all its complex interrelationships before a remedy can be considered; it is not easily understood why thin panels of granite may warp, though very little, similarly to marble despite a very different composition and geological history. On the other hand, the rapid decay of the brownstone popular in the eastern USA in recent years can be explained by the origin of the stone in desert environments in the geological past. The often unfavorable influence of chemical cleaners has aggrevated the problems even further.

After many years of study of stone decay it has become clear to me that only the general geologist with a broad background in the entire field of the earth sciences should be supervising the preservation and possible treatment of decaying stone structures. It is the geologist who understands the interactions of different types of stone with one another, the interaction of mortar with the stone, the catastrophic effect of deicing salts, and other phenomena. Forty years of teaching undergraduate students at the University of Notre Dame has given me much insight and broad knowledge of the field; teachers are learning through teaching (Dicendo discimus)!

Hope is expressed that this new book, Stone in Architecture: Properties, Durability, will fill a gap, providing updated basic information about stone, its properties and all aspects of decay, and the necessary background for the preservationist and anyone interested in stone and its natural beauty.


Notre Dame, Indiana                          E.M. WINKLER
August 1994

# Acknowledgements

The repeated request for my book, *Stone: Properties, Durability in Man's Environment*, published by Springer-Verlag, made me decide to prepare a new edition of this work, *Stone in Architecture: Properties, Durability*. This work was strongly supported by the John Skaggs and Mary L. Skaggs Foundation, Oakland, California, which has funded many international restoration projects. Their grants supported my research of stone decay and stone preservation for more than a decade. The present grant helps to update the two earlier editions of my book, paying for the preparation of all graphs and drawings, as well as the reproduction of the color plates by Springer-Verlag. For this generous contribution I am most grateful, especially to Mr. Philip M. Jelley, Secretary of the John Skaggs and Mary L. Skaggs foundation.

I would also like to thank the staff of the University of Notre Dame, Graphics Department of the College of Engineering, for their valuable assistance.

The manuscript was critically and expertly read by Mr. Victor VanBeuren, a science editor in New York City, to whom I am most grateful. Many practical suggestions for the improvement of illustrations and clarity came from my wife, Isolde, who also offered continuous assistance and encouragement during this undertaking.

# Contents

# 1 Rock and Stone

Rock is the basic building material of the earth's crust and the original building material used by man to serve as protection from the severe ice age and post-ice age raw climate; shelters were made of assembled field stones to form a basic house. Stone was later elaborately shaped to satisfy man's artistic expression.

The natural rock material is called "stone" when shaped to man's needs. We speak of building stone, decorative stone, sculptural stone, aggregate stone in the fabrication of concrete, rip-rap for shoreline protection against erosion, and others. It is only building stone, decorative stone and sculptural stone which are given attention in this book; stone properties and their decay also applies to aggregate stone and other types of stone. In contrast, "stone" in gemstones for bracelets and necklaces is not included.

All rocks are composed of one or more different minerals joined to one another in a more or less tight fabric which characterizes the stone and determines in part the physical and chemical properties, strength, color and durability. Most mineral properties may be found in basic texts of mineralogy. Appendix A summarizes common rock-forming minerals and the properties which determine the strength and durability of the stone.

Rocks are classified into three major groups based on their origin of formation: igneous, sedimentary, and metamorphic rocks. Figure 1.1 sketches the heat–pressure environment of formation of the three different rock groups.

1. Igneous or magmatic rocks are primarily crystallized from a fiery fluid silicate melt, taking place either deep below the earth's surface or at the surface. The fabric of these rocks depends on their environment of crystallization. Granites and basalts are the most common of this group. The stone industry summarizes coarse-grained rocks of this type "granites".
2. Sedimentary rocks, or layered rocks, are formed by the concentration of inorganic or organic debris of variable size and shape, deposited by mechanical means or by chemical precipitation. Conglomerate, sandstone, shale, limestone-marble, dolomite, travertine and onyx-marble are common sedimentary rocks in use.
3. Metamorphic rocks are igneous or sedimentary rocks recrystallized by the effect of temperature and pressure. Important rocks of this group

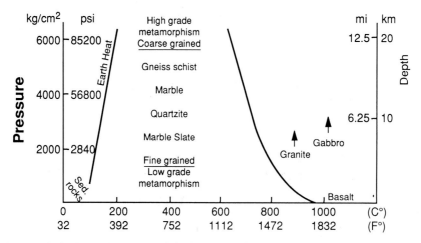

**Fig. 1.1.** Geological environments of the formation of rocks as to temperature and pressure

used in the stone industry are slate, crystalline marble, quartzite, greenstone and serpentine.

The following stone properties are important to the architect for proper selection:

Mineral composition: Most rock properties depend on the physical and chemical characteristics of the minerals; the larger the mineral constituents, the better the mineral components stand out and show their differential behavior.

Fabric: Fabric is the spatial orientation of mineral fragments within the rock.

Texture: Rock texture is the geometric aspect of the particles, e.g., size, shape, and arrangement. Texture overlaps with fabric; most geology texts combine fabric and texture into one term, "texture". Texture characterizes the sizes, shapes, and grain contacts, which differ greatly among the three major rock groups.

The structure of a rock mass consists of larger features which may not be visible on a small rock sample, such as bedding, folding, jointing, crossbedding, and others. Important major structural features are treated in Sections 3.1 and 3.3.

Color: Uniformity of color, color stability, and proper matching for interior and exterior applications are of great importance to the color- and fabric-conscious architect who is usually involved in the selection of stone for major buildings.

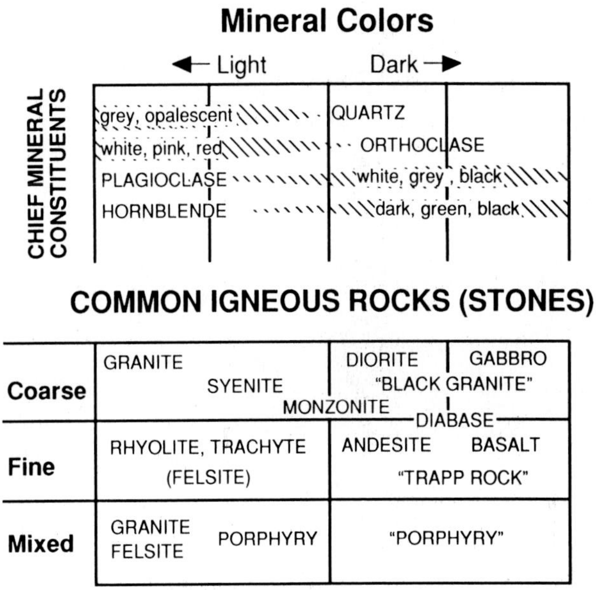

## Mineral Colors

← Light        Dark →

| CHIEF MINERAL CONSTITUENTS | | |
|---|---|---|
| grey, opalescent \\\\\\ | QUARTZ | |
| white, pink, red\\\\\\\ | ·· ORTHOCLASE | |
| PLAGIOCLASE \\\\\\\\ | \\white, grey, black\\\\ | |
| HORNBLENDE ········\\\\ | \\\dark, green, black\\\\ | |

## COMMON IGNEOUS ROCKS (STONES)

| | | | |
|---|---|---|---|
| **Coarse** | GRANITE     SYENITE | DIORITE   GABBRO "BLACK GRANITE" | |
| | MONZONITE | DIABASE | |
| **Fine** | RHYOLITE, TRACHYTE (FELSITE) | ANDESITE   BASALT "TRAPP ROCK" | |
| **Mixed** | GRANITE FELSITE   PORPHYRY | "PORPHYRY" | |

**Fig. 1.2.** Simplified classification of common igneous rocks, their mineral composition and mineral colors

## 1.1 Igneous Rocks

Igneous rocks, also called primary rocks, crystallize from a hot silicate melt, the magma. The term magmatic rocks is also used. If cooling progresses very slowly beneath the crust, crystallization is slow and the resulting crystals are generally coarse grained. Rocks formed in this environment are granite, syenite, diorite, and gabbro. The stone industry summarizes such rocks as granites (Fig. 1.2). If cooling takes place rapidly at or near the earth's surface, very fine-grained felsite, basalt, or porphyry will result. The classification of igneous rocks is mainly based upon the mineral content.

### 1.1.1 Igneous Textures

Igneous textures refer to the size and shape of the individual mineral grains.

Phaneritic or visible textures: almost all minerals that crystallize slowly under similar conditions will have almost equal grain sizes. Quartz grains and mica flakes, however, as a rule, are much smaller than the prismatic feldspars and hornblende. Most granites belong to this category (Fig. 1.3).

Aphanitic or nonvisible textures: Rapid cooling on or near the earth's surface causes crystallization of the entire rock mass homogeneously. The

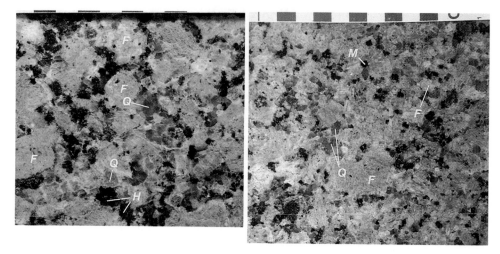

**Fig. 1.3.** Color and fabric of two different granites, determined by the color of the chief component, the feldspar. *F* Feldspar; *Q* quartz; *M* black mica, biotite; *H* black hornblende

**Fig. 1.4.** Porphyritic granites laid as a floor mosaic: large feldspar crystals quasi float in a finer groundmass of mostly feldspar, quartz, and dark hornblende. Exterior floor to the entrance of the Church of Loretto, St. Mary's College, Notre Dame, Indiana

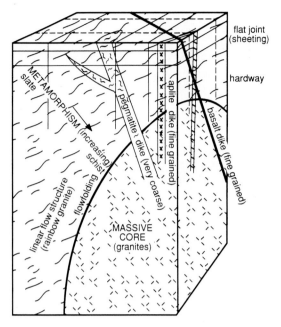

**Fig. 1.5.** Block model of an intrusive granitic body with its marginal phases, after removal of thousands of feet of overburden

minerals and their textural arrangement can be revealed only in a rock thin section. As a rule, finer grained rocks are harder and stronger than their coarser grained equivalents, felsites and basalts being the most important representatives.

Porphyritic: Large, well-developed crystals float in a fine-grained ground-mass or matrix. The contrast of the different crystal sizes reflects different environments of crystallization (Fig. 1.4). The remelting of a granite may show larger crystals damaged by melting attack. Other such mixed-crystal granites suggest two different environments of crystallization for the same rock: first, slowly, deep below the crust; then faster, after relocation of the magma or very rapid erosion of the covering rock masses. The process may involve millions of years. Porphyritic granites are always unique and are generally attractive as architectural and monumental stone.

Pegmatitic texture: The crytalline mass is usually large grained along veins and dikes, despite rapid cooling (Fig. 1.5). Gases act as catalysts and introduce rare elements, leading to the presence of rare minerals. Symmetric arrangement of crystals on both sides of the centerline of such dikes enhances the character even further. Though mostly ornamental, dikes create a twofold problem: cleavage of the large feldspar, mica, and hornblende crystals and, if the contact is sharp, a preferred separation of the coarser grained dike rock from the wall rock. A gradation of large minerals in the dike to the

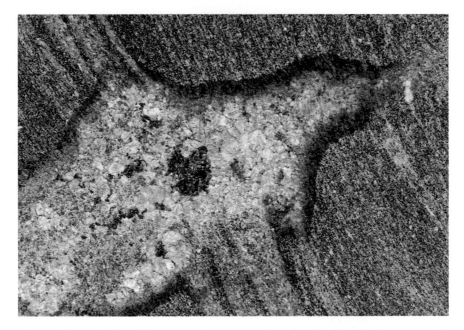

**Fig. 1.6.** Pegmatite dike with extension into granite with gneissic banding. The *brown margin* is caused by the reaction of the younger pegmatitic magma with the older gneissic rock. Tombstone of Brazilian granite

finer grained wall rock minimizes the danger of separation. Pegmatite veins are therefore considered as flaws by the stone producer for certain applications, such as on thin curtain walls (Fig. 1.6).

Aplitic texture: This is a light-colored, sugar-grained rock in veins that consist mostly of the light mineral components feldspar and quartz with no dark minerals present. They often resemble white calcite veins. The fine grains of the aplite veins, often finer than the grains of the wall rock, are the result of rapid cooling without the effect of the catalytic action of gases. Such contacts are potential lines of separation and water seepage (Fig. 1.6).

### 1.1.2 Igneous Structures

Granitic rocks are formed in intrusive bodies deep below the earth's surface, often with complex structures from high pressures. Convectional flow of the semi-liquid melt and interaction with the wall rock may complicate the texture and structure of granites. The exposed marginal zones of large intrusive bodies usually show such features. An idealized intrusive body is modeled in Fig. 1.5. The block diagram shows an intrusive body where a thick roof zone was removed by erosion. The diagram distinguishes between the massive core in the center, with random orientation of the minerals, and

**Fig. 1.7.** Coarse-grained group of orthoclase feldspars surrounded by swirls of banded "gneissic" rock. Rainbow granite, Cold Springs, Minnesota, on Chicago storefront

the flowfolding along the edge of the core, resembling metamorphic rock with local curling and twisting of the magma during a late intrusive phase. The Minnesota "rainbow granite" is an example (Fig. 1.7). True metamorphic rocks grade into the outer margin of such an intrusive mass, showing lower to higher grade metamorphism. The difference is not always readily discernible. Such marginal stone should be carefully tested for strength along flow structures. A mere blow with the hammer determines whether sufficient strength will keep the stone together and prevent later damage by weathering. Layers of mica flakes with its perfect cleavage may render the stone worthless, unless the layers are continuous and sufficiently spaced to split to flags for patios and walks.

### 1.1.3 Classification of Igneous Rocks

Igneous rocks range from almost white granites to black gabbros and basalts.
Acid magma, high in silica, crystallizes to rocks which are generally high in
orthoclase, quartz (pure crystallized silica), but low in dark minerals like
hornblende and the black mica, biotite. In contrast, basic magma is low in
silica and crystallizes to dark rocks that are high in gray to black plagioclase
and hornblende. The color of igneous rocks depends mainly on the color of
the prevailing feldspars which generally make up between 50 and 75% of the
total rock. Feldspar colors of granites and syenite may be white, pink, light
tan, or deep red. Diorites and gabbros, however, are medium gray to black
(Fig. 1.2). The quartz in granite may range from less than 10 to a maximum
of about 35%. The percentage depends on the abundance of silica in the
magma during crystallization. The percentage frequency of minerals may be
closely estimated from the mineral density chart of Fig. 1.8. The danger of
silicosis, decrease of strength, resistance to fire, and increased microcrack
porosity should be expected in granitic rocks with a high quartz content.
Syenites are quartz-free granites with a higher percentage of dark consituents

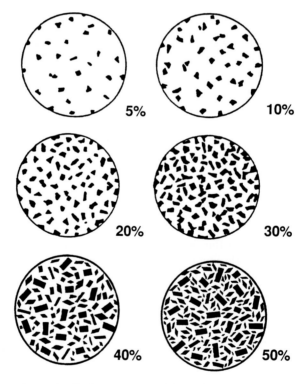

**Fig. 1.8.** Chart to estimate visually the percentage of dark minerals in a light-colored rock, or
of light-colored minerals in a dark rock. (Adapted from Terry and Chillingar 1955)

than granites. The rock classification chart of Fig. 1.2 gives an overview of the most important igneous rocks in the stone industry, their mineral composition, color, and texture. Commercial stone names are placed in quotation marks where they differ from the scientific terms. Granite is the most frequently used igneous rock because of its abundance and great variety of color and textures. Diorites and gabbros, often advertised as "black granites", are less frequently seen. Slag-like basaltic scoria are marketed as slag stones, mostly as accent in horticulture. Some slag basalts are filled with secondary mineral matter of a different color, mostly white or light green. Sponge-like, rusty-red basalt lava stones give storefronts in Mexico City a special character when used in rectangular blocks as dry masonry. Pale red, purple or olive-green porphyries with large, well-defined crystals have attracted sculptors in ancient Egypt and Greece. The same stone is marketed today under a variety of names, e.g., Porfido Rosso, and Porfido Verde Antico. Volcanic fragmentary rocks, such as ashes and cinders, are only useful when they are fused and well cemented.

## 1.2 Sedimentary Rocks

Sedimentary rocks, or layered rocks, are formed either by the accumulation of fragmentary rock material by streams, waves, or wind, or as organic accumulations and chemical precipitates. An important characteristic of sedimentary deposits is the flat, layered structure, known as bedding or stratification. The physical properties of sedimentary rocks depend primarily on the mineral composition, texture, fabric, structure, pore spaces, and grain cement. Sedimentary rocks are classified by their constituents and methods of formation (Figs. 1.9 & 1.10).

Fragmentary or clastic sediments:
a) Coarse- to fine-grained residual rock fragments produced through mechanical disintegration, found in various sizes and shapes;

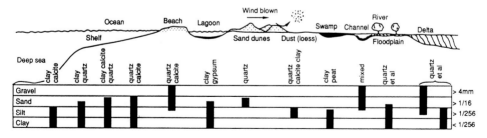

**Fig. 1.9.** Illustration of grain size distribution and the approximate mineral composition of different sedimentary environments

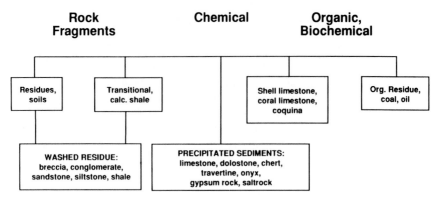

**Fig. 1.10.** Genetic relationship of sediments and sedimentary rocks

b) Fine to submicroscopic grain size: residual silt and clay-sized particles form the weathering end product after feldspars and some quartz.

Organic and biochemical sediments: The remains of animals and plants; many limestones, limestone-marbles, and dolomites belong to this group. Chemical precipitates from ocean water and brines: Some limestones, dolomites (dolostones), gypsumrock, and saltrock.

A mixture of these sedimentary materials is common in nature as many natural environments of deposition overlap. A simplified genetic relationship of common sediments is presented in Figs. 1.9 and 1.10.

### 1.2.1 Clastic Sediments

#### 1.2.1.1 Mineral Composition

Most minerals in clastic sediments were inherited from primary igneous rocks or from previously existing sedimentary or metamorphic rocks. Quartz, the hardest common mineral, survives mechanical abrasion in grinding processes on stream beds or on beaches. Quartz is also resistant to chemical attack. Carbonate fragments are often attacked by transporting undersaturated waters. Feldspars and ferromagnesian silicates slowly weather to clay; quartz is singled out as the sole survivor of the process of mass transportation and weathering. Soft, elastic mica flakes may survive chemical attack and some abrasion during transport to join quartz and clay minerals as the residue (Fig. 1.10).

Clastic sedimentary textures consist of the following components: sorting, roundness, packing, and fabric.

**Table 1.1.** Origin and characteristics of fragmentary or clastic sediments (see Fig. 1.10 to supplement this table)

| Stone | Environment of formation | Sorting | Rounding | Color |
|-------|--------------------------|---------|----------|-------|
| Breccia | No transport | Poor | Angular | Uniform |
| Conglomerate | Stream, beach | Fair to good | Round | Mixed |
| Sandstone, coarse | Stream | Good | Angular | Yellow, brown |
| Arkosic sandstone (feldspathic) | Derived from granites | Fair to good | Angular | Yellow, brown |
| Sandstone, fine-grained | Beach, dunes | Good, excellent | Well-rounded | Yellow |
| Coquina (shellstone) | Ocean shelf | Good | Poor | Gray, buff |
| Siltstone (not used) | Mudflat, shelf | Grains invisible | | Gray shale |
| | Mudflats, shelf | Flaky clay minerals are not visible | | |

*Sorting* indicates the degree of similarity of grain sizes which reflect the transporting agent. A full understanding of the environment of deposition often helps to understand unusual and unexpected behavior during laboratory tests. Table 1.1 presents the genetic environment of clastic sediments. The table may also serve as an approximate model for grain-size distribution, mineral composition, and expected color of sediments. A well-sorted dune sand contrasts with unsorted glacial deposits deposited by the indiscriminate pickup of debris by the snowplow action of an advancing ice mass. Sand may also occur in the rock matrix lodged between larger fragments of breccias or conglomerates. The matrix should not be confused with the rock cement. The same kind of deposit may display quite different grain sizes within a very narrow area: alluvial fans and streams often alternate between coarse flood material and the fine sand and silt deposits of minimum water flow and transport capacity. Oxidizing or reducing conditions during the time of sedimentation determine the original color: red, brown, yellow, and ochre under oxidizing conditions, and black, gray, green, and bluish-green under reducing conditions. The original color of the sediment, however, may change later due to a change in the oxidizing–reducing properties of ground-water.

*Roundness* of grains expresses the degree of abrasion of sandgrains by the sharpness of the edges and corners, as the radius of curvature of several edges of a solid to the radius of curvature of the maximum inscribed circle. Bayly's visual roundness scale is used where rapid identification is desired (Fig. 1.11) (Bayly 1968).

*Packing* of fragmental grains is the nearness of the grains to one another, the relationship of the grains to the intergranular spaces, and spacings. Houseknecht (1987) pictures four different examples of grain-to-intergrain spacings which determine the amount of grain cement and the uncemented open spaces, the capacity for water absorption, potential moisture travel and

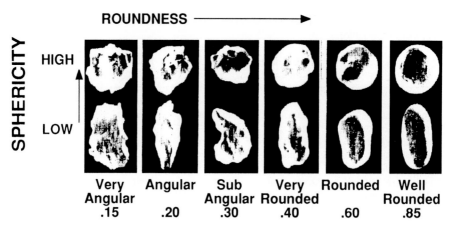

**Fig. 1.11.** Scale for the visual identification of roundness and sphericity of mineral grains. (Adapted from Bayly 1968)

the resistance to frost action, density, and strength. Figure 1.12 illustrates four different grain configurations accompanied by a graph and schematic diagram of the relationship between grain substance, grain coating or contact cement, the filler cement, and the remaining open pore spaces. Modified from Griffiths (1967), the distinction should be made between floating (case A), tangent (in part case B), complete (case C), and serrated by pressure solution (case D), which is rarely found in sandstone but is common in metamorphic rocks.

*Fabric* of sediments is the grain orientation – or lack of it; we therefore call it random (isotropic), or oriented by running water or wind (Fig. 1.13).

### 1.2.1.2 Cement of Clastic Sedimentary Rocks

Dissolved mineral matter, like calcium carbonate, silica, and ferrous iron, tends to travel through the pores of clastic sediments carried by groundwater and seeping rainwater; they precipitate as mostly calcite, quartz, and ferric hydroxide (common rust). The crystallinity is initially amorphous, which is metastable. It tends to crystallize by "aging", gradually dehydrating and strengthening the bonds with the grains. Inorganic stone consolidants follow a similar process. Three different types of natural cementation should be distinguished.

*Contact cement*, a thin film of precipitated mineral matter which coats the individual grains and cements the grains with each other at their contact points. The pore space of a contact-cemented clastic rock may be very large (Fig. 1.12, case A), unless the grain contacts are interlocked by high compression. The rock strength from the contact cement may be sufficient

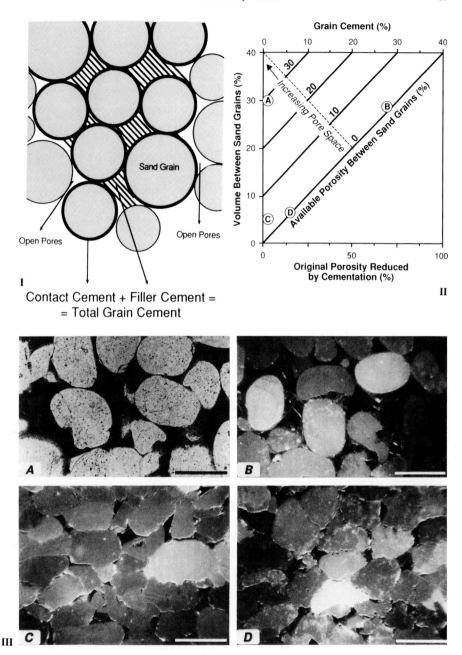

Contact Cement + Filler Cement =
= Total Grain Cement

**Fig. 1.12.** Relationship between the grain contact, grain cement, and the porosity of sandstones; adapted from Houseknecht. **I** Schematic section through a cemented sandstone, with grain contact and filler cement. **II** Different stages of cementation and filling of pore spaces. The letters *A–D* correlate with the actual photos of **III**. *A* Floating grains with 30% open pore spaces. *B* Floating grains with 29% pore spaces filled with cement. *C* Intergranular volume with serrated contacts with no cement. *D* Intergranular volume (6%) with open spaces occluded with quartz overgrowth; strongly serrated grain contacts by pressure solution

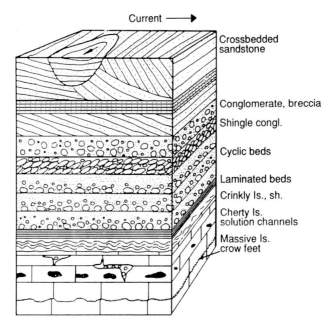

**Fig. 1.13.** Common structures of sedimentary rocks. *ls* Limestone; *sh* shale

provided that cementation is continuous and the cement is siliceous or calcitic. Rock fragments in stream beds and on beaches may be wrapped with a microscopic layer of clay, introduced when suspended mud or airborne dust temporarily filled streams and beaches. Clay-coated rock fragments appear sound in a dry climate but have caused failures when exposed to wetness and frost; the dry-to-wet strength ratio readily discloses such flaws.

*Pore cement* fills the interstices between the grains with or without the previous deposition of a contact cement coating the fragments. The composition of the pore or filler cement may be the same as the contact cement, or may be of a different age and composition. If there is a large volume, of pore cement in a floating grain, it is known as basal cement.

### 1.2.1.3 Composition of Rock Cement

The strength and durability of the rock depend on both the composition of the cement and the degree of cementation. The following mineral cements are recognized in commercial stone:

1. *Siliceous cement* may be introduced as finely crystalline quartz, as cryptocrystalline chert (chalcedony), or as amorphous opal. Rocks cemented

with silica may reach very high strength, especially if all the pores are filled with cement and no clay films coat the fragments. The rock tends to break through the grains, rather than around them, if the cement is precipitated around the fragments. The term orthoquartzite is used in contrast to metamorphic metaquartzite, whereby the prefix "ortho" or "meta" is omitted. Occasionally, the contact cement grows along the crystallographic orientation of the quartz grains; such sandstone tends to sparkle in the sun in one direction like fine-grained crystalline marble. Some Ohio sandstones display this feature. Sedimentary quartzites are distinguished from true metamorphic quartzites by their lack of uniform crystallinity. The dissolution of silica is possible under conditions of elevated temperature and pH. The environments of silica and calcite movement is discussed in Chapter 7.

2. *Carbonate* (calcite) *cement* is the dominant cement in sedimentary rocks. Thorough cementation with calcite or dolomite gives the rock sufficient strength and durability for most applications. The emplacement or removal of the carbonates of calcium or calcium–magnesium (dolomite) can change the porosity, strength, and durability with time.

3. *Ferric oxides and ferric hydroxides* are occasionally introduced as contact and/or pore cements. The iron may originate from black mica or hornblende, or from occasional grains of pyrite, the common sulfide of iron. Deposits of iron dust from nearby steel mills are not uncommon in urban-industrial areas. Cementation with hematite or limonite (natural rust) is usually incomplete. Iron travels readily in the ferrous-ferric form but becomes insoluble in most environments. Once the iron has changed to the insoluble ferric form the cement remains almost insoluble by the normal range of acid rainwater. This same iron oxide may form a fairly uniform patina-like buff to ochre-brown coating on light-colored granites and marbles.

4. *Clay cement* may also enter intergranular spaces, often as a secondary weathering product after feldspar. Careful testing should disclose such flaws.

5. *Feldspar cement* is occasionally found as extremely fine-grained filler cement in some sandstones. The poor performance record of some brownstones, popular in the eastern USA, may find its explanation in the very fine grains with a large specific surface to weathering attack to clay. Amorphous K or Na silicate, "natural waterglass", is believed to play a role in the original formation, possibly as with subsequent gradual crystallization to feldspars.

### 1.2.1.4 Structure of Clastic Sediments

Sedimentary rock structures refer to the type and thickness of bedding. The thickness of bedding largely depends on the clay and silica content of the rock substance. Silica in sediments increases the thickness at a constant lime content, whereas clay diminishes the thickness of beds from massive toward thin lamination. Short-lived floods or dust falls can introduce microscopic amounts of clay.

1. *Massive beds or banks* of sandstone and conglomerate exceed a thickness of 100 cm (3 ft). Massive sandstones are often crossbedded with siliceous or calcitic cement, illustrated in Fig. 1.14.

2. *Flagstone* beds are of medium slab thickness 3 to 6 cm thick. Such sandstone flags are common and are used for flooring and on walks and patios. The walking surface is usually sufficiently granular and often irregular to make it slip-proof when wet; ripple-marked wavy surfaces, worm tracks, and other markings add roughness and character to a walkway (Fig. 1.15).

3. *Crossbedding* is a sequence of horizontal strata alternating with uniformly inclined beds. The direction of the slope of inclined beds regularly changes (Fig. 1.16), e.g., in small deltas or wind-blown dune deposits. Channel fills are common in stream beds where an interruption or lowering of the water velocity can cause fine-grain deposition, resulting in cut-and-fill structures.

4. *Graded bedding* is a repetition of beds that show a coarse structure at the bottom and a fine-grained one at the top (Fig. 1.13). Such a sequence indicates seasonal stream deposition; coarse in springtime during flooding,

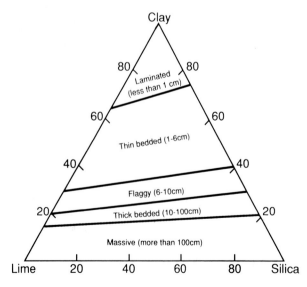

**Fig. 1.14.** Classification of bedding influenced by the proportion of lime to silica and to the clay content. (After Pettijohn 1957; Eicher 1968)

**Fig. 1.15.** Fine-grained sandstone, used as flagstone, with ripple marks. A triple wave cycle appears on the center slab. Scaling occurs where the ripples were cut into by erosion. Sandstone from Pennsylvania. Sidewalk on Canal Street, New Orleans, Louisiana

and fine in summer when less water is available. Graded bedding in stone blocks is uncommon.

5. *Shingle structure* is a crossbed-like arrangement of flat pebbles stacked like roof shingles on pebble beaches and in river beds. Though rare, shingle conglomerates give a distinct and directional pattern to a dimension stone.

### 1.2.1.5 Classification of Clastic Sediments

Classtic sediments are classified on the basis of their mean grain size and fabric.

**Fig. 1.16.** Crossbedded red sandstone (Buntsandstein). Pale green bands of ferrous iron were formed by reduction of the red matrix. Stone in building, Frankfurt, Germany

1. *Conglomerate* is composed of well-rounded and well-worn, often elongated, pebbles of various sizes and colors cemented together to give an often strong rock resembling concrete. The pebbles may float in a matrix of much finer granules and sand. Great distances of transport usually mix fragments of different composition and colors. As a result, conglomerates are priced as decorative stone. Differences of resistance to weathering show up after a few years of exposure to the atmosphere, especially when carbonate rocks are mixed with more resistant silicate rocks. Differential loss of polish and pitting of the polished stone surface are usually the result.

2. *Breccia* or *breche* is a coarse- to medium-grained fragmentary rock with angular, unworn fragments deposited near their source. Angular breches in a variety of colors and fragment sizes are commonly seen as decorative stone. Breches may also be found along faults as a result of mechanical crushing and fragmentation, as seen in Fig. 1.17.

3. *Sandstone* is composed of sand-sized grains, mostly quartz or feldspar, but also calcite, mica, and hornblende, with a variety of grain shapes and grain cements. A few common commercial sandstone varieties are discussed in the following.

*Quartzite* (orthoquartzite) is a sandstone or conglomerate that is composed of quartz. The cementation with silica is so prevalent that the rock breaks across grains instead of around them. Recrystallized, metamorphosed quartz sandstone is also called a quartzite (meta-quartzite), where the grains

**Fig. 1.17.** Breccia of pink Italian limestone. Crushing occurred along a fault zone, leaving angular fragments cemented with lithified residual lateritic soil. The stone is a popular decorative limestone marble for walls and floors

are no longer visible. Both types of quartzite are so dense and hard that working this stone appears impractical. The danger of silicosis demands wet drilling when working in the quarry and mill.

*Bluestone* is a hard, feldspathic gray sandstone which is often used for flooring.

*Brownstone* is a sandstone of distinctly brown or reddish-brown color, popular in the eastern USA. The depositional environment, a hot desert floor surrounded by mountains, supports the growth of feldspathic grain cement. The very fine grain size of the cement tends to weather to clay rapidly. The poor performance of this stone in humid climates is not surprising.

*Freestone* is a sandstone which splits with equal ease into any desired direction and dresses easily due to incomplete cementation of the sand grains.

## 1.2.2 Chemical Sediments and Evaporites

Chemical sediments are the result of precipitation from an aqueous solution. Our oceans are the chemical sink of all minerals freed by the weathering of rocks on land. These minerals may be redeposited by precipitation through limited interaction of various minerals with one another, or by evaporation of the solvent water. The classification chart of sediments in Fig. 1.10 draws lines between the chemical sediments, evaporites, and biochemical sediments. In nature, however, one type grades into another. Limestones and dolomites are usually impure, often gradating into pure mudstones. A practical classification of common clay–lime mixtures is reproduced in Fig. 1.18.

### 1.2.2.1 Textures of Chemical Sediments

*Fine-grained*: The grains are mostly invisible to the naked eye. The texture is microcrystalline. Very fine-grained homogeneous limestones are often used for lithoprinting. The term lithographic limestone is then used.

*Crystalline*: A coarsely crystalline limestone or gypsum rock recrystallized during lithification to larger grain size. This is a first step toward a true metamorphic crystalline marble. The difference may be readily identified by the presence of fossil fragments and warm iron-derived colors instead of cold whites or grays. The tan-to-pink Holston marble of Tennessee (Tennessee marble) is a popular stone of this category.

*Oolitic*, or roe-shaped and -sized, limestones often show a finely crystalline habit with concentric shells of calcite interspersed. This indicates deposition in warm shallow water. Oolitic textures are usually discernible with the naked eye. Some varieties of the Indiana limestone and Portland stone of England are oolitic.

*Skeletal* or *fossiliferous* limestone is composed primarily of the hard shell fragments of invertebrates. This material is usually well sorted and cross-bedded. Indiana limestone is a good example (see Figs. 7.10 & 7.11). Solution features tend to follow coarser layers with greater porosity; fossil shells are much more resistant to dissolution than the secondary fibrous and looser calcitic grain cement, standing out as an often noticeable relief on weathered stone surfaces. Fine-grained Indiana limestone develops a rough surface within a few years when exposed to rain.

### 1.2.2.2 Structures of Chemical Sediments

*Bedding* characterizes all sediments. The position of clay partings determines the position of the bedding planes.

*Crinkly beds* are limestones with local crenulation, found with some thin-bedded evaporite limestones.

*Ripple marks* and *mudcracks* may be observed on the surfaces of fine-grained clastics and carbonates (Fig. 1.16).

*Stylolites* and *crow's feet* are irregular wavy seams which resemble graphs of daily temperatures. Stylolites are the result of interstratal pressure solution in carbonate rocks, during which the solution causes the upper strata to collapse into the lower undissolved bed, filling the gap with the insoluble residual clay that tends to thicken near the apex of the zigzag lines. Weathering along such clay-filled seams and the complexity of this prominent feature is discussed in Section 7.2.

### 1.2.2.3 Classification of Chemical Sediments and Evaporites

*Limestone* is principally composed of calcium carbonate with less than 5% clay impurities (Fig. 1.18). Limestones are subject to large variations in color and composition, attributed to their origin and history of lithification. Limestones low in clay are called commercial limestone-marble. They take a good polish and are strong and durable.

*Magnesian (dolomitic) limestone* contains between 5 and not more than 40% magnesium carbonate, denser than pure limestone. Dolomite with more than 40% magnesium carbonate often appears finely crystalline and porous due to extensive recrystallization from low-magnesian carbonate rock.

*Dolomite* with more than 40% magnesium carbonate appears often finely crystalline and porous due to extensive recrystallization from low-magnesian carbonate rock. Dolomites commonly are Paleozoic in age, whereas limestones are more recent. Dolomites are generally harder, heav-

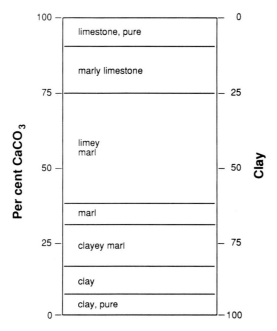

**Fig. 1.18.** Nomenclature of limestone–clay mixtures. (After Pettijohn 1957)

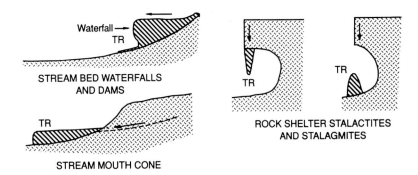

**Fig. 1.19.** Principles of the formation of travertine, tufa, and onyx-marble. (After Viles and Goudie 1990)

ier, and more brittle and appear more fractured than limestones. Their use is the same as that for limestones.

*Travertine* is a banded carbonate of rapidly changing textures and a product of chemical precipitation from some cold, but mostly hot, natural springs. Such deposits along streams and in caves are irregular in shape, yet the layers tend to be parallel. They can be massive at the foot of hills, from where calcite-saturated water drains and redeposits the carbonate by evaporation in semi-humid to semi-arid climates; this usually occurs with the help of vegetation because water plants of different kinds extract the $CO_2$, thereby reducing the solubility of the water. Figure 1.19 shows typical accumulations of commercial quantity travertine. Dense varieties are favorites for walls and flooring; looser textured varieties are favored for decorative wall panels, usually set in book or diamond match. The evenly cream-colored Italian or Roman travertine from Tivoli was the favorite building stone for ancient Rome. The best-known buildings are the Colosseum and Bernini's Colonnades to St. Peter's Cathedral. Colored varieties with coarser bedding are found in Colorado and Utah.

*Onyx-marble* is a recrystallized variety of travertine, very fine-grained and translucent with soft pastel colors, mostly caused by iron of different concentrations and states of oxidation, such as green, cream, and golden brown; these often follow bedding planes and cracks (Fig. 1.20). The excellent light transmission of thin panels of onyx has invited the use of such panels as substitutes for glass windows with protection from the bright sunlight in the Mediterranean area. Galla Placidia's Mausoleum in Ravenna, and the windows in the cupola of the Hagia Sofia in Istanbul are the best-known examples of historical use. The light reflection–refraction combination gives an almost opal-like effect when used on walls and in bathrooms. The extensive onyx deposits near Puebla, Mexico, and in various areas in Italy, Turkey, Iran, and Pakistan have produced tiles of exotic beauty. Large wall panels of Mexican onyx spread diffuse warm light into the

**Fig. 1.20.** Onyx-marble as "geological picture windows" in a residential living room. The stone panels are ⅞ in. thick and are polished on both sides for better light transmission. The stone panels were cut against the bedding planes, along which ferric iron has travelled. Stone from Tres Enriques, Baja California, Mexico

College of Surgeons' Library reading room, in Chicago, during daylight, and inside lighting causes a similar reverse effect at night.

*Shell limestone* contains numerous fossil seashells with a variety of sizes and shapes, well cemented, and oriented by ocean currents; the shells appear as cross sections after fabrication into slabs. Occasional cross sections of white calcite shells in a deep black Blue Belgian limestone-marble have provided special effect to generations of architects. The shell imprints as casts and molds in the buff Austin Shell stone of Texas form large vugs in the remaining matrix. Cross sections of large shells, snail-like ammonites, and other fossils are known to be found in a large variety of decorative stones, giving character and contrast.

*Coral limestone* as building stones and more frequently as decorative stones often comprise entire coral colonies on a single stone slab. They are generally white globular coral sections with a differently colored matrix and show a variety of skeletal orientations. Figure 1.21 shows interior decorative stone with corals, quarried in the Alpine area. Coral limestone of recent reef areas now elevated above sea level are frequently quarried for building stone and broken stone. The degree of cementation and induration is usually high enough to make it a solid building stone. The stone is a favorite in southern Florida, where it is known as Key Largo limestone, and in the South Pacific.

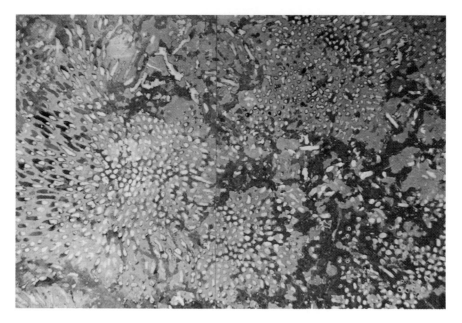

**Fig. 1.21.** Coral limestone-marble. White coral branches are surrounded and filled in with red limestone mud. Lobby of Vienna Western Railroad Station, Austria

*Chalk and chalky limestone* is a soft, poorly lithified, white to gray evaporite of very fine grain size, great porosity, and low strength. Gradations are found toward a dense and sound limestone.

*Chert* may occur as nodules, flat lenticular bodies or as wavy, often, continuous, bands in limestones. The concentration of silica has probably taken place during lithification. Chert tends to stand out in dimension stone by its gray color. Organic concentrations often color chert dark to black, known as flint. The resistance of chert to weathering, compared with the surrounding soluble limestone, can develop an undesirable surface relief. The presence of chert in softer limestone can cause considerable trouble in drilling operations if soft-rock drill bits unexpectedly hit the hard chert. Flying silica dust in quarries and mills requires maximum ventilation and wet drilling to avoid silicosis (see Chap. 12).

*Gypsum rock* is a densely crystalline rock composed of the soft calcium sulfate gypsum; it ranges in color from white to pink, and is translucent when cut to thin plates called alabaster. The alabaster windows of Galla Placidia's Mausoleum in Ravenna spread a warm pink color as the only source of indoor lighting; these stone windows are actually translucent, harder onyx-marble. The softness of alabaster and its high solubility in water does not recommend exterior application of any kind. Anhydrite of similar composition and appearance is the anhydrous sulfate of calcium. The stone is as hard as calcite but as soluble as gypsum. Both rock types are found associated with other evaporite deposits, like salt deposits. Despite great solubility, very coarse-grained gypsum as dimension stone blocks have held up well throughout the centuries at the base of the Asinelli Towers in Bologna.

## 1.3 Metamorphic Rocks

Igneous and sedimentary rocks subjected to increased pressure and temperature by burial beneath accumulating sediments or crustal movements tend to adjust to their new environment by recrystallization to more stable mineral phases. New minerals develop which chemically resemble the parent material but are more stable under the greater heat and pressure. The process is very slow and often associated with intense folding. The heat caused by the vicinity of igneous bodies or by geothermal activity at great depths develops plastic conditions for the rock masses. These allow molecular migration from one layer to the next and beyond. These processes last millions of years. The concentration of minerals along bands causes a gneissic structure or foliation. High-temperature minerals are stable and may not lose their identity. The original massive unoriented mineral arrangement of igneous rocks changes to a distinctly gneissic structure. The degree of metamorphic intensity generally influences the size of the individual mineral grains: low metamorphic zones lead to the formation of slates, fine-grained marbles, and granite-gneisses, whereas higher metamorphic processes form schists and coarse-grained marbles. Table 1.2 presents common metamorphic rocks, their mineral composition, and their possible parent-rock material. Not all metamorphic rocks are of commercial stone quality. Schists rich in

**Table 1.2.** Origin and mineral composition of metamorphic rocks

| Low metamorphic | High metamorphic | Mineral content | Parent rock |
|---|---|---|---|
| Granite-gneiss | Same | Feldspar, quartz, hornblende | Granite, arkosic sandstone, conglomerate |
| Hornblendeschist, greenstone | Same | Hornblende, mica, plagioclase | Diorite, gabbro, shale, porphyry, basalt |
| Quartzite, fine | Coarse | Quartz (mica) | Quartz-sandstone, conglomerate |
| Marble, fine | Coarse | Calcite (quartz, mica) | Limestone (impure) |
| Dolomite-marble, fine | Coarse | Dolomite | Dolostone, dolomitic limestone |
| Slate | Mica, schist | Mica (quartz, calcite) | Clay, shale |

mica tend to split unevenly and are therefore unfit. Only certain gneisses, greenstones, quartzites, and slates can be utilized.

### 1.3.1 Textures of Metamorphic Rocks

All metamorphic minerals are tightly interlocked, resulting in a serrated texture (Fig. 1.22). The common textures are as follows:

1. *Microcrystalline*, or slaty, texture is composed of submicroscopic mica flakes, some quartz, and other minerals which tend to arrange in parallel fashion.
2. *Granular*, or granoblastic, texture is composed of minerals of near-equal grain size; found in marbles, quartzites, and many gneisses.
3. *Porphyroblastic* texture is equivalent to the igneous porphyritic texture. Flaky and prismatic minerals often wrap around the larger crystals, the porphyroblasts.

### 1.3.2 Structures of Metamorphic Rocks

Metamorphic structures reflect the character and the degree of metamorphism from igneous or sedimentary rocks.

**Fig. 1.22.** Thin section through a slab of crystalline marble, viewed through a petrographic microscope. The interlocking of the mineral grains, about 2 mm in size, give the rock strength. Laas marble, Italy

**Fig. 1.23.** Calcite crystal with distinctly visible lines of cleavage, compared with a rhombohedral cleavage piece. Calcite cleavage in three dimensions can cause easy chipping in coarse-grained varieties

1. *Massive*: granular rock with massive appearance and without visual orientation of the grains, e.g., some marbles, quartzites, and serpentines known as Verde Antique.

2. *Banded*: nearly parallel bands of minerals of different texture and color; characteristic of gneisses.

3. *Foliation*: splitting of the rock into thin sheets by the presence of tabular or prismatic minerals.

4. *Lineation*: streaking on a foliated surface, which indicates movement caused by hard minerals enclosed in a generally softer matrix. Foliated rocks are often lineated.

5. *Platy cleavage*: foliation results from even thickness by the parallel arrangement of microscopically sized mica. Rock cleavage should not be confused with mineral cleavage (Fig. 1.23). Cleavage can be irregular and incomplete because of interruption by small nodes and pits of quartz grains or other minerals. Imperfect cleavage also may result in a plumose pattern on the cleavage planes of slate (see Sect. 3.1). Slate with such markings adds character when used on walks, patios, or as thin wall cladding.

6. *Ribbon slates*: some slates show the original bedding of sedimentary structures cutting across the slaty cleavage. Ribbon slates can be quite ornamental if marked by darker lines or different colors. Ribbons may also

**Fig. 1.24.** Banded or "ribbon" slate. The bands are sedimentary relict structures. Ferric or red iron oxide is occasionally reduced to greenish ferrous iron oxide. Bands are at a steep angle to the slaty cleavage. New York red slate

be considered as flaws when they form lines of weakness or greater hardness (Fig. 1.24).

7. *Folding*: some gneisses and marbles are tightly folded and formed by plastic or semi-plastic flow during metamorphism. Such folds are generally not lines of weakness and may give a desirable ornamental accent to the stone (Figs. 1.25 & 3.22).

### 1.3.3 Mineral Composition of Metamorphic Rocks

The minerals of metamorphic rocks are either inherited from igneous or sedimentary rocks, such as feldspars, quartz, micas, hornblende, calcite, and dolomite, or are newly formed during the process of metamorphism, such as mica, sericite, chlorite, garnet, and hornblende. Pigments of metamorphic rocks are discussed in Section 4.5.

### 1.3.4 Classification of Metamorphic Rocks

*Gneiss* is a rock with a granitic look but a strong parallel arrangement of the grains with mineral concentrations along bands. A true gneissic structure should have a good compressive strength perpendicular to the long axis of the mineral grains but a less favorable strength parallel to the grain. Gneisses are often classified as granites.

*Schist* is similar to gneiss but with a thinner banding (schistosity) than gneiss. The decrease in feldspars and quartz, and the increased presence of mica and hornblende enable rock splitting upon impact. The rock may have enough strength to find application as rough slabs for wall veneers and flagstones, provided that the prismatic minerals are more abundant than the flaky micas.

*Marble* (metamorphic crystalline) is the most frequently used metamorphic rock for thin wall panels. When uniform, it is highly priced; when fine grained, it is excellent for sculptures. Interlocking grains of mostly calcite or dolomite (Fig. 1.22) give these rocks maximum density and strength with a minimum pore space. The color may be a cold white, gray, pink, or green, often occurring in streaks and bands. The colors of true metamorphic marbles are "cold" compared with the soft, warm pastel tones of sedimentary limestone-marbles. The warmer colors are caused by ferrous and ferric iron compounds as well as organic admixtures. Pure white fine-grained massive

**Fig. 1.25.** Brecciated Carrara marble, metamorphosed twice. Angular fragments accumulated at the base of a bluff were cemented and later metamorphosed a second time. The fragments were elongated and slightly folded by plastic flow. Base of Pieta statue, Sacred Heart Church, Notre Dame, Indiana

marble is often marketed as monumental, memorial, or statuary stone. Coarser grained varieties, however, are less favorable. The perfect cleavage of calcite in all three directions may become an obstacle to precision sculptural work. White, sound marble has attracted famous sculptors throughout human history. Phidias and Praxiteles worked with Pentelic marble in Greece. Michelangelo, Pisano, and other sculptors of the Renaissance, more recently Mestrovich, worked with Carrara marble. In the USA, fine-grained Vermont marble and gray to pink coarse-grained Georgia marble are popular exterior veneer stones.

*Serpentine*, often called serpentine-marble, or sometimes Verde Antique, is composed of the mineral serpentine, a magnesium silicate. The stone is frequently used for its beautiful green color. White veinlets of calcite or magnesite, the pure magnesium carbonate, enhance the character of the stone. The presence of magnesite, however, may become a serious problem in sulfate-polluted atmospheres because the magnesium carbonate recrystallizes readily to the highly water-soluble sulfate of magnesium. The great durability of the mineral serpentine permits both interior and exterior applications. Calcite veins in serpentine show the loss of polish of the calcite in exterior paneling.

*Greenstones* are mostly metamorphosed basic igneous rocks of the basalt type, with the green color derived from green hornblende and chlorite. Greenstone may be used like any other dimension stone when not foliated.

*Quartzite* (metaquartzite) is a recrystallized quartz rock with interlocking grains of quartz, occasionally joined by a few flakes of mica. The great hardness of this rock causes problems in quarrying and finishing. Silica-

**Table 1.3.** Summary of the major common rock types used in the stone industry

| Igneous rocks | | | |
|---|---|---|---|
| | *Intrusive* | *Intermediate* | *Extrusive* |
| Light colored (acid) | Granite | | Felsite |
| | Syenite | Porphyry | Felsite |
| Bark colored (basic) | Diorite | | Andesite |
| | Gabbro | | Basalt |
| Sedimentary rocks | | | |
| | *Coarse* | *Medium* | *Fine* |
| Clastic | Conglomerate | Sandstone | Siltstone |
| | Breccia | | Shale |
| Chemical | | Dolomite, gyprock | Organic limestone |
| Biochemical | Coquina, shell limestone | Shell limestone | |
| Metamorphic rocks | | | |
| | *Coarse* | *Medium* | *Fine* |
| From igneous | Gneiss, schist | Gneiss, schist | Serpentine |
| From sediments | Gneiss, schist, marble | Gneiss, schist, marble, quartzite | Slate, marble, marble, quartzite |

cemented sedimentary quartz sandstones may also be called quartzites. The distinction is only minor as both rock types are almost entirely composed of quartz and are of equal hardness and strength. The color of quartzite is generally white or light yellow, but may also be red in the presence of iron oxides and greenish in the presence of mica. Free silica dust in quartzite quarries should be considered a health hazard which may cause silicosis (see Chap. 12).

*Slate* is a microgranular metamorphic rock resulting from low-grade metamorphism. The perfect slaty cleavage is of variable thickness. Great strength and durability are typical of good-quality slates that are low in calcite. Incompletely metamorphosed slate may show both the slaty cleavage and still the original bedding planes, which are preserved as "ribbons" (Fig. 1.24). The ribbons may be either lines of weakness or zones of greater hardness compared with the slaty matrix. Where they are exposed to heavy foot traffic, the ribbons tend to form a relief. Some slates develop feather-like markings, starting at the point of chisel impact. Such feathers, nodes, and small pits on cleavage surfaces add character if used for walks and other exterior surfaces (Fig. 3.5).

# References

Bayly B (1968) Introduction to petrology. Prentice Hall, New York, 371 pp

Eicher DL (1968) Geologic time. Prentice Hall, New York, 149 pp

Griffiths JC (1967) Scientific methods in analysis of sediments. McGraw Hill, New York, 508 pp

Houseknecht DW (1987) Assessing the relative importance of compaction processes and cementation to reduction of porosity in sandstones. Am Assoc Pet Geol Bull 71(6):633–643

Pettijohn RJ (1957) Sedimentary rocks, 2nd edn. Harper and Row, New York, 718 pp

Powers MC (1953) A new roundness scale for sedimentary particles. J Sediment Petrol 23: 117–119

Terry RD, Chillingar GV (1955) Summary concerning some additional aids in studying sedimentary formations. J Sediment Petrol 25:229–234

Viles HA, Goudie AS (1990) Reconnaissance studies of the tufa deposits of the Napier Range, N.W. Australia. Earth Surface Processes and Landforms 15:425–443

# 2 Physical Properties of Stone

Stone is a heterogeneous substance characterized by a wide range of mineral compositions, textures, and rock structures. Consequently, the physical and chemical properties and the resulting durability are quite variable. The suitability of a stone for a given building can be easily tested in the laboratory. Although some tests are expensive and consume considerable amounts of rock material, others are simple, inexpensive, and nondestructive.

## 2.1 Rock Pores and Porosity

Porosity is the ratio of the volume of pore space to the total volume in percent. The definition by Walker et al. (1969) of porosity for concrete also applies to stone, with the apparent or effective porosity

$$n = V_v/V_b \times 100,$$

where n is the total effective porosity, $V_v$ the total pore volume determined by the mercury intrusion at up to 1000 atm, and $V_b$ the bulk dry volume of the sample in $cm^3$.

The average porosity comprises most of the diameters as the sum porosity between all the diameters (Bayly 1968).

The modulus is the cumulative sum of the penetration of fluid per gram of rock, of predetermined diameters, expressed in $cm^3/g$.

The porosity number p is calculated as

$$p = B - A/V \times 100,$$

where A is the weight of the dried specimen, B the weight of the soaked specimen, and V the total volume of the sample.

The porosity increases rapidly in clastic sediments with decreasing grain size, thus clays may have a porosity well over 50%, according to Bayly (1968). The water in micropores is not readily available and cannot migrate.

The porosity of rock and stone helps to determine the strength and durability, but also permits estimates of the content of moisture and travel through the masonry. A full understanding of the size and character of such channel ways is therefore important (see Sects. 6.3–6.5).

**Table 2.1.** Porosity of sandstones and carbonates (Choquette and Prey 1970)

| Properties | Sandstone | Carbonates |
|---|---|---|
| Original porosity | | |
|    Uncemented | 25–40% | 40–70% |
|    Cemented | 15–30% | 5–15% |
| Pore sizes | Related to particle size | Little related to particle size and sorting |
| Pore shape | By particle shape | Varies greatly |
| Semiquantitative evaluation | Easy | Variable |
| Porosity-permeability relation | Consistent: depends on sorting, particle size | Vugs, channels unpredictable |

### 2.1.1 Origin of Rock Pores

#### 2.1.1.1 Igneous Rocks

The tight fit of the mineral fabric in all igneous rocks formed under great pressure and temperature permits very little open pore space. The anomalous behavior of the quartz grains, contracting more than half their volume during the process of cooling, causes extensive cracking across and around the quartz grains. This leads to a microcrack porosity of about 1.0% of the total volume, twice the amount of quartz-free igneous rocks (see Figs. 10.2–10.4).

#### 2.1.1.2 Sedimentary Rocks

The porosity and pore space distribution is subject to great variation. A nearly unlimited variety of pore sizes and shapes are characteristic for sedimentary rocks, where they serve as aquifers and oil reservoirs. Choquette and Pray (1970) distinguish between megapores (256–0.062 mm diameter) and micropores (0.062–0.0001 mm). Table 2.1 compares the kind and sources of pores in sedimentary rocks. Houseknecht (1987) sketches the relationship of grain packing, grain shape, and pore filling with available pore spaces (see Fig. 1.12).

#### 2.1.1.3 Metamorphic Rocks

Mostly foliated or oriented rocks are densely packed by the influence of pressure and heat during formation with a minimum of available pore spaces. Such rocks were not exposed to heat high enough to cause total melting. Quartz in gneisses does therefore not increase the porosity (see

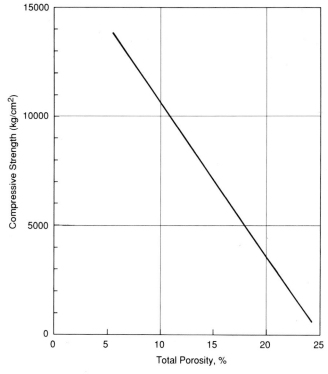

**Fig. 2.1.** Comparison of porosity with uniaxial compressive strength. Adapted from Fitzner (1988)

**Table 2.2.** Porosity and bulk density ranges of some common stones (Farmer 1968; for more densities, see Daly et al. 1966)

| Rock (stone) | Bulk density | | Porosity (%) |
|---|---|---|---|
| | gm/cm$^3$ | lbs/ft$^3$ | |
| Granites | 2.6–2.7 | 162–168 | 0.5–1.5 |
| Gabbro (black granite) | 3.0–3.1 | 187–193 | 0.1–0.2 |
| Rhyolite (felsite) | 2.4–2.6 | 150–162 | 4.0–4.6 |
| Andesite (felsite) | 2.2–2.3 | 137–144 | 10.0–15.0 |
| Basalt | 2.8–2.9 | 175–181 | 0.1–1.0 |
| Sandstone | 2.0–2.6 | 125–162 | 5.0–25.0 |
| Shale | 2.0–2.4 | 125–150 | 10.0–30.0 |
| Limestone | 2.2–2.6 | 137–162 | 5.0–20.0 |
| Dolomite | 2.5–2.6 | 156–162 | 1.0–5.0 |
| Gneiss | 2.9–3.0 | 181–187 | 0.5–1.5 |
| Marble | 2.6–2.7 | 162–168 | 0.5–2.0 |
| Quartzite | 2.65 | 165 | 0.1–0.5 |
| Slate | 2.6–2.7 | 162–168 | 0.1–0.5 |

Sect. 10.4). Interlocking calcite grains of crystalline marble have a pore space of 0.5% or less.

While the total porosity figure permits the correlation with other physical parameters, e.g., compressive strength (Fig. 2.1) and ultrasound velocities, the porosity distribution permits a full evaluation of progressive weathering toward the stone surface (Fitzner 1988) (see Fig. 6.8), the progressive effect of frost and salt action in test cycles, and the diminishing effectiveness of stone consolidants with time (see Fig. 11.7). Instrumentation for the determination of the porosity is complex and expensive; porosity determined by the pore radius method appears to be the most important variable in the evaluation of stone properties and the effectiveness of stone consolidants. Table 2.2 gives the bulk density and the total porosity for a variety of common stones.

## 2.2 Water Sorption

Absorption and adsorption are terms often used for the same property, sorption. The terms are readily confused. The occurrence of liquid water in stone capillaries is outlined in Sections 6.2–6.5 and illustrated in Figs. 6.2–6.7. Capillaries smaller than $0.1 \mu m$ do not absorb water. Vacuum soaking fills most capillaries. "Absorption" is the take-up, assimilation, or incorporation of gases in liquids, whereas "adsorption" is of molecules or ions in solution, or solutions of different concentration becoming more concentrated upon contact. The percentage of water (ab)sorption into stone capillaries by weight is determined by

$$B - A/B \times 100,$$

where A is the weight of the oven-dried specimen at $105\,°C$, and B the weight of the specimen after immersion in water. Favorable stones range between 3 and 15% water. Water sorption can be by submersion or forced by vacuum into available capillaries, leaving dry the capillaries of smaller sizes in which oriented (structured) water blocks the passage ways. Such sizes can absorb moisture from humid air (see Figs. 2.16, 6.2 & 6.3.). The small size of the water molecule does not readily allow the study in narrow capillaries.

## 2.3 Bulk Specific Gravity

The bulk or apparent specific gravity is the ratio of the mass of an equal volume of water at a specified temperature. The bulk specific gravity is calculated as $A/A - B$; A is the weight of the dried specimen, and B the weight when suspended and soaked in water. The bulk specific gravity multiplied by 62.4 gives the unit weight per cubic foot of dry stone. Table 2.2 summarizes bulk specific gravities for a variety of stones used in the building trade.

## 2.4 Rock Hardness

The hardness of a mineral or rock is its resistance to indentation or scratching. This is an important factor to estimate the resistance to mechanical wear and the workability of a stone in the quarry and mill. The hardness is consistent on fresh mineral surfaces and is easily tested in the field and in the laboratory. We distinguish between the scratch hardness, the indentation hardness, the abrasion hardness, the rebound hardness, and the impact hardness. The complex mineral association and variability in rocks makes a correlation of the physical parameters very difficult. Rock drilling deals essentially with impact and abrasion hardness. The important hardnesses in stone use are discussed in the following.

### 2.4.1 Scratch Hardness

Scratch hardness is the ability of one solid to be scratched by another harder solid, as a complicated function of the elastic, plastic, and frictional properties of a mineral surface. Friedrich von Mohs noticed in 1822 that a certain mineral from different sources had the same hardness. The quantitative hardness is expressed as the minimum weight required to produce a visible scratch. The scratch hardness is the oldest practical hardness scale, created by von Mohs in 1822 as an arbitrary scale based on ten common minerals (Fig. 2.2; Table 2.3).

**Table 2.3.** Common minerals in the Mohs hardness scale

| Hardness | Mineral | Tool (hardness) | Other minerals |
|---|---|---|---|
| 1 | Talc | Thumbnail (2½) | Graphite |
| 2 | Gypsum | Thumbnail (2½) | Epsomite, kaolinite (copper) |
| 3 | Calcite | Copper coin (3¼) | Anhydrite (low-carbon iron, copper) |
| 4 | (Fluorite)* | Glass plate (5½) | Dolomite, siderite |
| 5 | (Apatite)* | Glass plate (5½) | Magnetite, goethite |
| 6 | Feldspar | Good knife (6¼) | Hornblende, augite, hematite, pyrite |
| 7 | Quartz | Corundum (9) | (Some ceramics, high-carbon tool steel) |
| 8 | Topaz* | – | |
| 9 | Corundum* | – | (Silicon carbide, carborundom) |
| 10 | Diamond* | – | (Tri-boron nitrite) |

* Minerals unimportant in the stone industry; ( ) nonmineral substances.

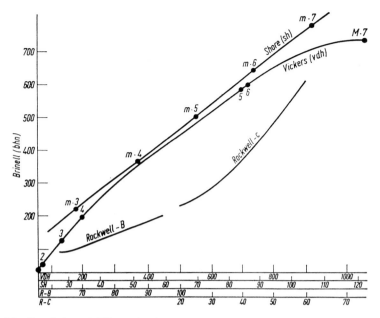

**Fig. 2.2.** Correlation of different hardness scales for minerals and metals. *M*, Mohs hardness scale, original arbitrary scale; *m*, Mohs hardness scale, quantified; *VDH*, Vickers diamond hardness; *SH*, Shore rebound hardness; *R-B*, Rockwell B hardness (indentation); *R-C*, Rockwell C hardness (indentation). Data compiled by author from metallurgic laboratories

The fingernail and a pocket knife with a hard blade suffice in the field to distinguish between graphite, gypsum, calcite, and quartz. Figure 2.2 shows the arbitrary character of the Mohs hardness scale; it was meant to be a simple guideline using available testing tools. Mohs' scale, beginning with gypsum and ending with diamond, established a nonquantitative scale between 1 and 10 with which to compare other minerals. The greater hardnesses are spaced further apart than the soft values, with quartz being the hardest common rock-forming mineral. Despite 170 years of the Mohs hardness scale no other workable field scale has been developed. Most gemstones are of hardness H = 6 or harder. The stone industry is not interested in gemstones. The common minerals critical in stone are kaolinite (clay), gypsum, calcite, dolomite, hornblende, and quartz; these can be distinguished on the basis of their hardness. Proctor (1970) compares the original Mohs hardness to common rocks in order to estimate the boring rates of tunneling machines. Figure 2.3 gives the average Mohs hardness ranges for some common rocks.

Granites and gneisses are controlled by the quartzes and feldspars, with a hardness range between 6 and 7.

Basalts and felsites are determined by feldspars and hornblende, with a range of 5 to 6½.

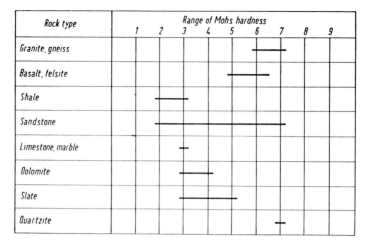

**Fig. 2.3.** Rock hardness guide to common rocks, based on the Mohs hardness scale. After Proctor (1970)

Shales range from 2 to 3, or even less when wet. These rocks are too soft and unsuitable for the stone industry.

Sandstone has the greatest range of 3 to 7. The degree of cementation and hardness of the grain cement determine the total hardness of a sandstone.

Limestones, close to H = 3 when well lithified (uniformly cemented), can be deceivingly soft, e.g., chalk. Admixtures of clay minerals can soften such a limestone when wet to less than H = 2, easily scratched with the thumbnail, for durability. Nodules of chert (H = 7) are also detectable with a pocket knife. They will not be cut or scratched by the knife. The test is quick and a reliable signal for whether further testing should be carried out.

The hardness of crystalline marble is determined by the percentage of calcite. All clay will have changed to micaceous minerals which, though soft, will not swell. The presence of graphite, if significant, may soften the stone locally along dark bands.

Dolostones range from H = 3.5 to 4, depending on the ratio of calcite to dolomite. Like in pure limestones, the presence of finely distributed clay or clay in concentrations can lower the hardness when wet.

Slate, a very dense rock mostly composed of mica and quartz, ranges from H = 3 to 5, depending on the mineral proportion.

Quartzite, both sedimentary and metamorphic, is the hardest rock with H = 7. A rock of less than H = 7 should not be called a quartzite.

### 2.4.2 Indentation Hardness

Indentation hardness is the permanent indentation with a loaded indenter of a brittle material as a sphere, cone, or pyramid. The load and size of

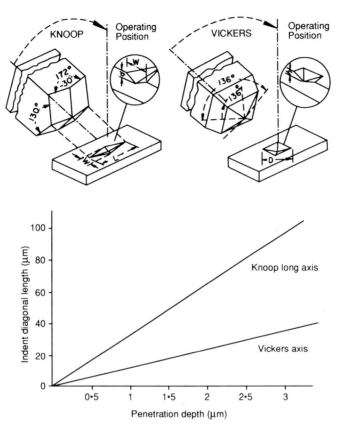

**Fig. 2.4.** Comparison of the Knoop and Vickers diamond indentation hardness, and the relationship of the diagonal length and the depth of penetration. (Adapted from McColm 1990)

indentation determines the hardness. Data of rock mechanics often include the rebound hardness. Indentation hardness is quantitative, and was developed by the metallurgist. Known as the Brinell, Vickers, Knoop, and Shore scleroscope hardness, all are restricted to use in the laboratory with expensive equipment. Residual stress fields develop in the vicinity of an indentation near the surface of the material; they have a compressive radial and a tangential tensile component that decreases as the third power of the distance from the indentation center (Fig. 2.4) (McColm 1990).

a) Brinell hardness, $H_b$, has long been in use to test metallic and ceramic materials. A hard spherical steel indenter of diameter D presses onto a smooth mineral surface under a load W (kg). The mean chordal diameter D of the resultant indentation gives d (mm). The Brinell hardness number is then calculated:

$$H_b = W/D \, [D - (D^2 - d^2)] \, kg/mm^2.$$

b) Vickers diamond hardness, $H_v$, is similar to the Brinell hardness. A pyramidal diamond indenter presses into the metal or mineral surface greater than the projected area of the indentation by the ratio $1:0.9272$; the Vickers hardness is calculated as:

$$H_v = 0.9272 \times W/\text{area of indentation},$$

which equals $1.8544\,W/d^2$, in $kg/mm^2$. The Vickers hardness is similar to the Brinell hardness at low hardness numbers. Generally, the indentation hardness increases with decreasing main grain size (Brace 1961).

c) Knoop hardness, $H_k$, is distinguished from Vickers hardness by the shape of the diamond indenter. Figure 2.4 compares the $H_v$ with the $H_k$. The long diagonal length does not change by possible elastic recovery in metals and is therefore preferred for metal hardness determination. McColm (1990) also shows the relationship of the chordal length with the depth of penetration. The Knoop hardness therefore appears more sensitive and accurate.

d) Rockwell hardness is based on the depth of penetration with a preload of $10\,kg$; $H_r$ consists of the $H_r$-B and the $H_r$-C, which is first applied to the surface and retained for the main test, recorded for a further $150\,kg$. Preloading the specimen surface eliminates errors through elastic recovery. A spherical diamond indenter is used with a hemispherical tip for harder minerals.

e) Shore scleroscope hardness and Schmidt hardness are both based on rebound. Scleroscope hardness is a measure of the elastic properties of a rock or concrete as the rebound of a steel ball or a diamond-pointed hammer is dropped vertically onto the test surface. The rebound is measured on an arbitrary scale of 120 divisions. Crushing decreases the rebound energy and the rebound height by an amount equal to the crushing energy and the energy absorbed by the rock surface and the instrument. The simplicity of the instrument and operation permits many readings to be taken in a short time both in the laboratory and in the field. Rebound can also be measured on vertical surfaces, e.g., on a masonry wall, with specially designed instruments. The Schmidt hardness, similar to the Shore hardness and also based on rebound, is obtained with a spring-loaded Schmidt test hammer designed to estimate the strength of concrete or stone in place on a wall. A low impact energy of $0.54$ ft-lb with the L-type hammer is favored for testing stone as soft rock tends to break on impact at greater energies. Good correlation can be obtained for the Shore and Schmidt impact hardness with the compressive strength, described by Deere (1968). The graph in Fig. 2.5 separates rocks with different dry unit weights.

f) Drilling hardness, or the drillability, is of importance to the stone industry to separate the stone blocks in the quarries for further manufacturing. Rotary and percussion or impact drilling are the common techniques.

Rotary drilling, with a diamond-studded drill crown mounted on a hollow tube, can obtain solid continuous rock cores for detailed study and

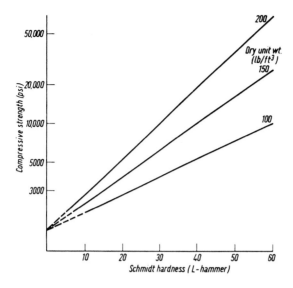

**Fig. 2.5.** Relationship between the Schmidt impact hardness, uniaxial compressive strength, and the dry unit weight. (Adapted from Deere 1968)

**Table 2.4.** Mineral hardnesses compared with the Mohs scale. (Compiled from Tertsch 1949)

|            | Mohs (M) | Vickers ($H_V$) | Knoop ($H_K$) | Abrasive ($H_A$) | Drilling |
|------------|----------|-----------------|---------------|------------------|----------|
| Talc       | 1        | 47              | –             | 0.003            | –        |
| Gypsum     | 2        | 60              | 46–54         | 1.25             | 8.3      |
| Calcite    | 3        | 105–136         | 75–120        | 4.5              | 50       |
| Fluorite   | 4        | 175–200         | 139–152       | 5.0              | 143      |
| Orthoclase | 6        | 714             | 560           | 37               | 4 665    |
| Quartz     | 7        | 1103–1260       | 666–902       | 120              | 7 648    |
| Corundum   | 9        | 2085            | 1 700–2 200   | 1 000            | 188 808  |
| Diamond    | 10       | –               | 80 000        | 140 000          | –        |

testing. The drilling hardness of minerals was brought to attention by Tertsch (1949). A rotary diamond point drills a depression into the mineral surface at 6 to 7 rev/s. The drilling hardness consists of the number of revolutions necessary to cut a hole 10 $\mu$m deep. The drilling hardness is proportional to other hardness values (see Table 2.4).

Percussion drilling fragments the rock at the bottom of the hole with hardened steel ridges at the bottom of the drill rod; a center hole permits water to flush the cuttings out of the hole. Important mechanical variables are involved, such as rock hardness, impact toughness, drill-bit sharpness, and drillshaft pressure; the drilling progress is thus difficult to predict.

g) Abrasion hardness or abrasive strength is the resistance of a mineral or rock to abrasive wear, developed by Tertsch (1949). A given quantity of abrasive is used without renewal with an abrasion time of 8 min. Today, the abrasion hardness is based on the Dorry abrasive resistance, $H_A$, with the simple equation,

$$H_A = 10\,G\,(2000 + W_s)/2000\,W_a,$$

the reciprocal of the volume of material abraded, multiplied by 10; the superimposed weight of the specimen of 2000 g plus the weight of the specimen ($W_s$), the loss of weight during the grinding operation, $W_a$, and the bulk specific gravity, G (ASTM C-241-90). Dense rock, such as fine-grained limestone, records higher values than coarse-grained varieties, because the grains loosen more easily along the larger interface area. This test method covers the determination of the abrasion resistance for all types of stones for floors, steps, and similar uses where the wear is caused by the

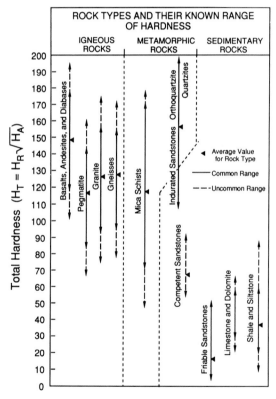

**Fig. 2.6.** Ranges of the total hardness for some common rocks (compare with Fig. 2.3) (adapted from Tarkoy 1981). $H_T$ Total hardness; $H_R$ Schmidt rebound hardness (L-type); $H_A$ abrasion hardness, on NX-size discs 0.5-cm thick and abraded for 400 revolutions on each side of the disc; weight loss in g

abrasion of foot traffic. Quartz sandstone with a weaker calcitic cement records $H_a$ values closer to calcite than to quartz grains which are readily twisted out of the stone fabric during grinding. Thus, the true abrasion hardness does not depend on the average mineral hardness and on the strength of the interface bond.

h) Total hardness, $H_T$, was developed from the abrasion hardness and Schmidt impact hardness, to predict the rock performance for estimates of the boring progress. $H_T$, hinged on the impact and the abrasion hardnesses, is well applicable to the general workability of stone, both in the quarry, in the mill, and on the site. The total hardness ranges of common rock types is given in Fig. 2.6. In general, rock hardness decreases with increase of grain size.

i) Brittleness of minerals and rocks may show a brittle or a plastic behavior upon impact. Calcite and quartz are considered brittle, whereas metallic copper is plastic, malleable, or ductile.

## 2.5 Compressive Strength

The unconfined or uniaxial compressive strength is the load per unit area under which a block fails by shear or splitting. It is an important parameter in rock testing for comparison with other rock strength values. Rocks are unconfined near the earth's surface and on buildings, whereas they are confined deep beneath the earth's surface; triaxial compressive strength reflects rock strength in mountain building, deep rock foundations, and tunnels. The ASTM standards C-170-90 recommend uniaxial strength tests both parallel and perpendicular to the chief stress on 2-in. (5-cm) cubes, dry and water soaked (Fig. 2.7). The compressive strength is calculated as

$$C = W/A,$$

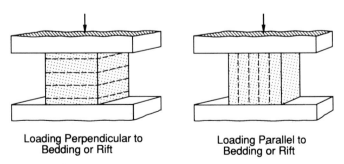

Loading Perpendicular to        Loading Parallel to
Bedding or Rift                 Bedding or Rift

**Fig. 2.7.** Position of test cube for uniaxial compressive strength, with the load perpendicular or parallel to the rift of the sample (ASTM C-170)

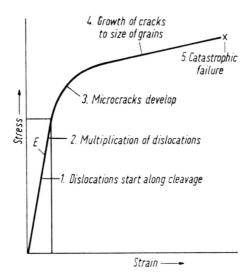

**Fig. 2.8.** Stress–strain relationship in compression and tension. The modulus of elasticity, E, is the ratio of stress to strain. The progressive destruction of the test specimens is marked on the diagram. (After Conrad and Sujata 1960)

| Term | Psi | kg/cm | Rock materials |
|---|---|---|---|
| Very weak | <1000 | <70 | |
| Weak | 1000 - 3000 | 70 - 200 | |
| Medium strong | 3000 - 10,000 | 200 - 700 | |
| Strong | 10,000 - 20,000 | 700 - 1400 | |
| Very strong | > 20,000 | >1400 | |

**Fig. 2.9.** Strength classification of important stones on the basis of uniaxial compressive strength. (After Hawkes and Mellor 1970)

where C is the compressive strength of the specimen, W the load in lbs on the specimen at failure, and A the calculated area of the bearing surface in square inches. The rate of loading should not exceed 100 psi/s to permit internal adjustment of the stone fabric.

Uniaxial compression and tension develop a basic stress–strain pattern (Fig. 2.8) from which the modulus of elasticity, E, is calculated from the stress–strain curve. Dislocation takes place in the first two stages along cleavage planes. The strain increases rapidly when microcracks begin to develop, leading to the growth of cracks to the size of the grains. Catastrophic failure is the final stage of compression or tension.

The uniaxial compressive strength of rock has been used as the general index for rock strength. Bedded and foliated rocks record different rock strengths depending on the angle of bedding or banding. The compressive strength increases with decreasing grain size in homogeneous rocks, like rock salt, crystalline marble, metamorphic quartzites, and others. Hawkes and Mellor (1970) give a general strength classification (Fig. 2.9).

The compressive strength is also related to the tensile strength and the modulus of rupture.

## 2.6 Tensile Strength

Tensile strength is the degree of coherence of the stone to resist the pulling force; this depends both on the strength of the mineral grains and on the cement or interface area from one mineral grain to the adjacent grain. Fracture starts at the grain boundaries and detachment follows as the tension continues (Brace 1964). Tensile strength for the stone industry is replaced by the easier "hoop test" developed by Hardy et al. (1971); a doughnut-shaped ring of the test specimen is broken by hydraulic inside expansion. Tensile strength is often replaced with the modulus of rupture, a combination of compression and tension.

## 2.7 Modulus of Rupture

The modulus of rupture or flexural strength is the resistance of a rock slab to bending or flexure (ASTM C-99-52, 1990). Wind stresses, snow loads, stacking loads, and warped stone slabs on buildings may bring forth such stresses. The flexural test is centerpoint loaded by a knife edge; it should be performed parallel and perpendicular to bedding or banding, with the knife edge being as long as the specimen is wide. Figure 2.10A sketches the principle of the setup; a disc, with a neutral centerline, a-a, has compression in the upper zone and tension below the centerline. Prim and Wittmann (1985) have developed a tester similar in principle to ASTM Standard C-880 (Fig. 2.10B). The load transfers onto a 90-mm test disc, which replaces the 10-in.-long test bar of ASTM C-880. The disc is large enough and the apparatus sensitive enough to test both the modulus of rupture and the modulus of elasticity. Winkler (1986) uses discs that are only 1½ in. in diameter, sliced from drill cores at the desired orientation. The use of small discs permits a large number of specimens to be cored and sliced from a block only half the size of a masonry brick; these are sufficient to test dry-to-

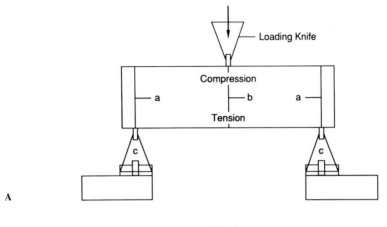

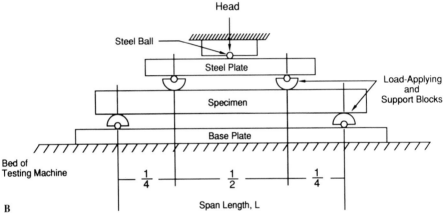

**Fig. 2.10. A** Modulus of rupture (ASTM C-99). The bar or disc rests on the edges *c*, with an effective span *a-a*; the center-line is through *b*. There is compression above the line and tension below.

$$R = 3Wl/2bd^2,$$

where R is the modulus of rupture in psi or MPa, W the breaking load in lb or kg, l the length of span from *a* to *a*, b the width of the specimen, and d the thickness of the specimen. **B** Principle of flexural strength (ASTM C-880)

wet strength ratios, or the efficiency of consolidants. The small size of the specimens, however, requires rocks of medium- to fine-grain size. A correlation of the modulus of rupture with water absorption is attempted for marbles of different grain size, fine-grained Carrara marble and coarse-grained Georgia marble (Fig. 2.11; Winkler 1986).

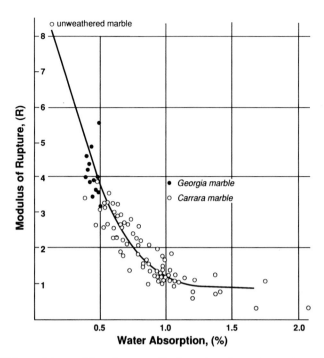

**Fig. 2.11.** Values of the modulus of rupture, R, depend on the moisture absorption. (Winkler 1986)

## 2.8 Modulus of Elasticity

Rock properties are governed by the reaction of the rock to the forces acting on it; such forces induce a state of stress which results in deformation, the state of strain. The relationship between stress and strain is the basis for the modulus of elasticity, E. The linear stress–strain relationship is the modulus of elasticity expressed as $E = s/e$, where s is the stress, and e the rate of strain. The limit of elastic deformation is the strength in brittle material (Fig. 2.8). Fine-grained rocks are quasielastic, whereas coarse-grained, still

**Table 2.5.** E values for some rocks at zero load (Farmer 1968)

| Rock | E (kg/cm$^2$) $\times$ 10$^5$ | Rock | E (kg/cm$^2$) $\times$ 10$^5$ |
|---|---|---|---|
| Granite | 2–6 | Microgranite | 3–8 |
| Syenite | 6–8 | Biorite | 7–10 |
| Gabbro | 7–11 | Basalt | 6–10 |
| Sandstone | 0.3–8 | Shale | 1–3.5 |
| Limestone | 1–8 | Dolomite | 4–8.4 |

cohesive rocks are semielastic. Coarse-grained rocks with high porosity are nonelastic. E values are given in Table 2.5.

A linear correlation is possible with flexural strength and the modulus of rupture.

## 2.9 Thermal Properties of Minerals and Rocks

1. Rock temperatures in deserts and on stone buildings can reach 60 °C, or higher on dark stone surfaces. Temperatures recorded on black and white surfaces can serve as a basis for surface temperatures, for instance a black "granite" and a white marble surface (Cullen 1963). Figure 2.12 compares solar temperatures on a roof from 8 o'clock in the morning to 8 o'clock in the evening. Nightly cooling is reflected in the morning temperatures. McGreevey and Smith (1982) give a number of maximum surface stone temperatures from various desert areas of the world. Black basalt and dark sandstones of the Sahara recorded almost 80 °C. Different stone surface temperatures on buildings in urban Belfast, Ireland, according to McGreevey (1985) are a function of the albedo, the maximum temperature, and the

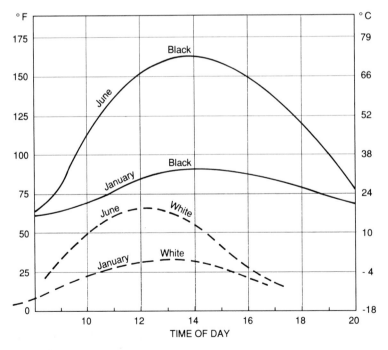

**Fig. 2.12.** Surface temperatures of a black and a white roof surface for different hours of the day in January and June, near Washington DC (39°N). (Adapted from data from Cullen 1963)

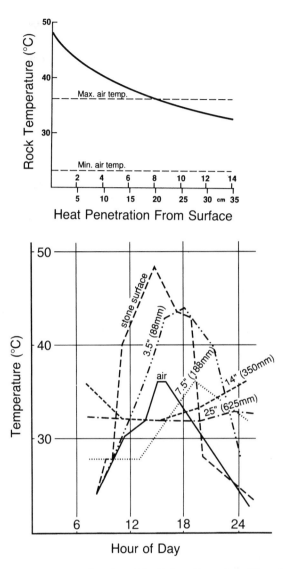

**Fig. 2.13.** Stone temperatures as a function of depth from the surface. The values are compared with the ambient air temperature. (After Roth 1965)

thermal conductivity in close correlation with one another. The peak surface temperatures depend on the conductivity and on the albedo. Measurements taken 5 cm below the stone surface depend much on the thermal conductivity. A temperature profile for various stone depths through a gray granodiorite boulder was recorded with implanted thermistors, recorded at various depths. The temperature maxima shift toward afternoon hours with in-

**Table 2.6.** Urban temperature, albedo, and thermal properties (McGreevey 1985)

|           | Max. temperature | | Albedo (%) | Thermal conductivity |
|-----------|-------|----------|------------|----------------------|
|           | Surface | 5 cm below |          | (W/m/°C)             |
| Chalk     | 24.2  | 23.6     | 25         | 1.72                 |
| Granite   | 27.4  | 25.8     | 18         | 1.65                 |
| Sandstone | 28.5  | 26.5     | 12         | 1.05                 |
| Basalt    | 29.2  | 26.3     | 12         | 0.96                 |

**Table 2.7.** Thermal conductivity and expansion. (Data compiled from Skinner 1966)

| Rock      | Conductivity ($10^{-3}$ cal/ cm °C linear, %) | Thermal expansion | Critical mineral     |
|-----------|-----------------------------|-------------------|----------------------|
| Granites  | 5–7                         | $8 \times 10^{-6}$ | Quartz (5–35%)       |
| Basalts   | 5                           | 5.4               | –                    |
| Limestone | 8                           | 8                 | Calcite (90–100%)    |
| Dolomite  | 7–8                         |                   |                      |
| Sandstone | 5–9                         | 10                | Quartz (50–100%)     |
| Quartzite | 14–15                       | 11                | Quartz (100%)        |
| Marble    | 5–6                         | 7                 | Calcite (90–100%)    |
| Slate     | 4–5                         | 9                 | Quartz (5–40%)       |

creasing depth and flatten to a depth of about 7.5 in. The stone surface temperature is almost twice the ambient air temperature (Fig. 2.13; Roth 1965). Albedo and surface temperatures are easy to measure; the albedo is measured with a photographic spot meter against the Kodak Neutral Gray Card (Sect. 4.1) with known reflectance. Table 2.6 presents some data.

2. Thermal conductivity reflects the insulating capacity of stone, an important property for building material. The thermal conductivity can be calculated with the formula

$$K = q \cdot L/A \ (t_1 - t_2) \ [ASTM \ C\text{-}177],$$

where K is thermal conductivity in Btu/h/ft$^2$/°F; q is rate of heat flow in Btu/h; L is thickness of the specimen in in., measured along a path normal to the isothermal surface; A is area of the isothermal surface, in ft$^2$; $t_1$ is temperature of the hot surface in °F; and $t_2$ is temperature of the cold surface.

The thermal conductivity (Table 2.7) and heat retention of dense rocks is higher than of porous rocks. The values are important for flooring, thin wall panels, and for heat storage when stone is considered for passive solar heating of buildings.

3. Thermal and moisture expansion (Table 2.8) of stone and concrete has long been known from observations in the field and in the laboratory.

**Table 2.8.** Moisture expansion of some building stones (Hockman and Kessler 1950)

| Stone | Source | Expansion (%) |
|---|---|---|
| Biotite-granite | Barre, VT | 0.005 |
| Biotite-granite | Mt. Airy, NC | 0.002 |
| Gabbro | St. Peters, PA | 0.0056 |
| Basalt-porphyry | Columbia, WA | 0.0018 |
| Limestone | Bedford, IN | 0.0028 |
| Limestone, porous | Cedar Park, TX | 0.0015 |
| Marble, coarse crystalline | Georgia | 0.0025 |
| Marble, fine crystalline | Vermont | 0.0010 |
| Sandstone, quartz | Amherst, OH | 0.013 |
| Sandstone, quartz | Glenmont, OH | 0.010 |

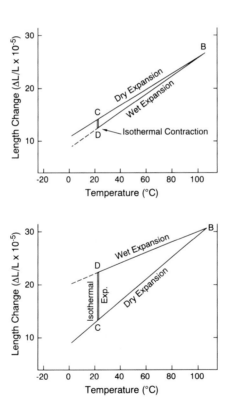

**Fig. 2.14.** Length change of stone slabs of dolomite as a function of water saturation and temperature. *Above* Wet expansion is less than dry expansion in sound dolomites. *Below* Wet expansion by far exceeds the dry expansion in dolomites with a poor durability record. (Hudec 1980)

Hockman and Kessler (1950) determined the thermal and moisture expansion of samples of 48 different granites. Expansion by mere immersion of the granite and basalt in water averages to 0.029%. The authors have observed nearly equal expansion by temperature and by moisture for granites, with the total expansion as the sum of moisture and thermal changes. A 100-ft course of granite is expected to expand 0.05 in. by wetting and 0.08 in. due to a 10 °C increase in temperature, a total of 0.13 in. Stone blocks in coping courses tend to move progressively toward the end of the course without returning to their original position; the resulting cracks fill with water, etc., and frost action may continue the destructive work.

Hudec (1980) compares the temperature–moisture expansion of durable, weathering-resistant carbonates with sorption-sensitive carbonates in his durability study of carbonate rocks (Fig. 2.14); while the difference between dry and wet expansion is minimal in sound stone (top graph), wet expansion exceeds the dry expansion considerably at a given temperature below 100 °C (lower graph). The presence of chert and alumina, mostly as clay substance, probably take the blame for such unfavorable behavior.

The European literature summarizes such behavior as "hygric expansion" (Felix 1983). The dry-to-wet strength ratio appears to be related with thermal–moisture expansion.

## 2.10 Dry-to-Wet Strength Ratio

The poor performance of sandstones of the Swiss and German Alpine Molasse, used since the Middle Ages for most historic buildings of Switzerland and southern Germany, challenged Tetmajer (1884) to compare the compressive strength of dry and watersoaked stone for suitability. Tetmajer's report has remained unnoticed for a century. Colback and Wiid (1965) plotted the uniaxial compressive strength against the moisture content of two rock types, both watersoaked and at various relative humidities (RH) of the ambient atmosphere. The reduction of strength after exposure to 15% RH and above is important for masonry exposed to high RH values in urban areas (Fig. 2.15) Michalopoulos and Triafilidis (1976) believed that the pore water softens the bonding strength, especially in the presence of clay minerals. The stone appears sound when dry, but when wet can soften to such an extent that it may fall apart. The great pore pressure of water in narrow capillaries can override softening, when free sorption is compared with sorption under vacuum, despite greater penetration of the moisture into the capillary system. The ratio between these values can serve as the blocking of large pores by stone consolidants. Dry-to-wet strength ratios can be plotted for the modulus of rupture, uniaxial compressive strength, flexural strength, and other factors. Figure 2.16 plots ratios for the approximate

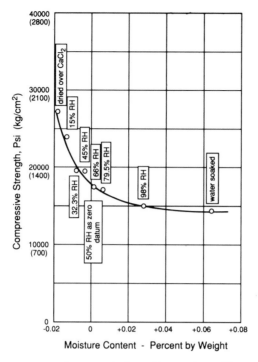

**Fig. 2.15.** Uniaxial compressive strength depending on the ambient relative humidity of the atmosphere (RH). The *curve* is drawn for a quartzose siltstone, with the 50% moisture content as the zero line. Curves of this kind should be established for each stone planned for use as cladding in areas of high RH. (Adapted from Colback and Wiid 1965)

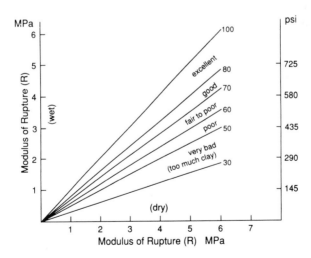

**Fig. 2.16.** Dry-to-wet strength ratios valid for most stone types. Evaluation of stone quality is based on lines of equal ratios. (Winkler 1986)

quality rating of the rock. The values can be readily obtained with small discs if a quick decision is needed for the potential use of a stone.

## 2.11 Ultrasound Travel in Stone

The travel of ultrasound waves is related to the travel of earthquake waves through rock masses, which cause mechanical stresses. In concrete, the sound velocity equals the square root of the modulus of elasticity, the E value, over the raw density of concrete. Figure 3.25 distinguishes the three basic types of waves during an earthquake; this is discussed in detail in Section 3.4. All three wave types are emitted from an earthquake focus or from a wave transmitter through a transducer in a similar way. The wave

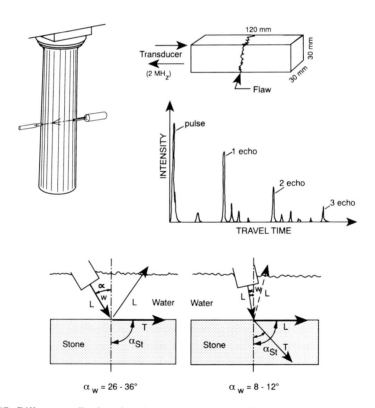

**Fig. 2.17.** Different applications for ultrasound wave travel through stone. *Top block* Wave reflection through a stone block with use of a single sender–receiver transducer. Bouncing of waves shows echoes with decreasing intensity. *Column* Wave travel across a stone column or block, with sender–receiver transducers in opposite position. *Bottom stone blocks* Transducers angled to the stone block surface can reflect or refract longitudinal (*L*) and transverse (*T*) waves, tested under water. Compiled from Mamillan (1976) and Volkwein (1982)

length ranges from 50 kHz (50 000 cycles/s or 50 kilohertz) to over 1 MHz (1 million cycles/s or 1 megahertz). Commercial instruments have been in use for decades to generate such pulses as flaw detectors in order to test the soundness of concrete pillars and walls for cracks. The complexity and heterogeneity of stone fabrics, however, has complicated and often confused transmitted wave images. Bundled waves are easy to apply to fine-grained and homogeneous steel, but are almost impossible to obtain in stone with multiple dispersing grain surfaces. Proper contact of the transducer with a rough stone surface is the major problem, and is mitigated using Vaseline jellies or water baths. An angled position of the transducer to the stone surface can control the use to transmit either longitudinal or transverse waves (Fig. 2.17). More information can be obtained from Krautkraemer and Krautkraemer (1990). Fine-grained, dense stone varieties respond to greater cycles in the near 1 MHz range, while porous sandstones may respond to only near 25 kHz.

The following useful techniques have been developed.

1. Reflection: Fine-grained stone varieties can be tested by wave reflection (Fig. 2.17) in a stone block with a single transducer by multiple reflection echoes. Internal refractions can obscure the wave patterns. Very fine-grained stone varieties only can be successfully tested by reflection, with which velocity and possible flaws can be located.

2. Longitudinal waves (L-waves): The degree of weakening of the signal is a function of the modulus of rupture (Fig. 2.18), R, in fine-grained limestone. The travel velocity at a given wavelength can also indicate the soundness of a stone. This transmission-velocity method is used to test crystalline marble (Fig. 2.19).

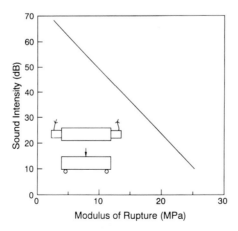

**Fig. 2.18.** Relationship between ultrasound intensity and the modulus of rupture, R. Figures taken of dense limestone-marble (Royal Perlato), with $U_L$ = 6090 m/s. (Volkwein 1982)

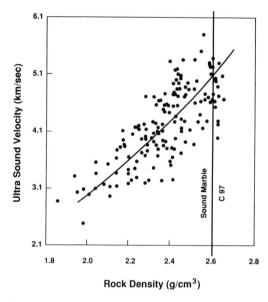

**Fig. 2.19.** Density of crystalline marble versus ultrasound velocity. The density of sound marble is set at 2.6 g/cm³ and 5.1 km/s. (Adapted from Gaviglio 1989)

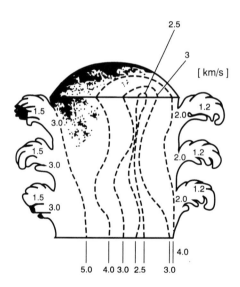

**Fig. 2.20.** Soundness of Carrara marble, characterized by ultrasound wave travel velocity. Ornaments on the capital are most vulnerable to decay. The low velocity of the inner core may be erroneous. Capital of Sansoussi, Potsdam, Germany. (Koehler 1988)

3. Transverse waves (T-waves): Traveling at half the wavelength of the L-waves, T-waves can also indicate the strength of a stone. They are absorbed by water-filled pore systems. The angle of the transducer to the stone surface affects the transmission. Angles between 26 and 36° favor T-waves, while reflecting L-waves; those between 8 and 12° favor L-waves with T-waves being refracted through the stone.

The following stone properties have been studied by the nondestructive sonic technique.

1. Rock soundness as a function of the wave travel velocity is often applied to marble statues (Mamillan 1976; Koehler 1988). Crystalline marble and dense fine-grained limestones respond to wave travel at 0.5 to 1 MHz wavelength. Koehler (1988) contoured a marble capital for ultrasound velocities as a function of the degree of stone decay with the lowest readings for ornaments most exposed to weathering (Fig. 2.20). Also, Tarkoy's (1981) chart of the rippability of some rock types as a function of the wave velocity compares well with the rock soundness (Fig. 2.21). While porosity decreases the wave velocity, moisture saturation increases it (Fig. 2.22). The wavelength and type of transducer must be matched for correct readings.

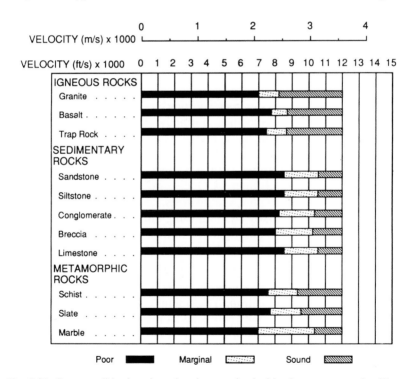

**Fig. 2.21.** Stone qualities based on the ultrasound velocities for common rocks. The graph is adapted from a rippability chart as a guide for tunneling boring machines. (Adapted from Tarkoy 1981)

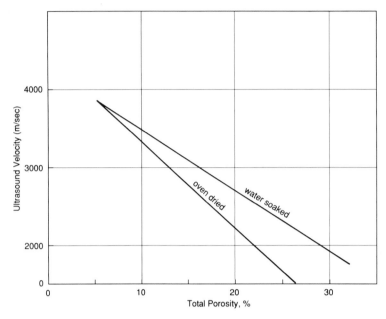

**Fig. 2.22.** Ultrasound travel velocity as a function of the total porosity, dry and water soaked. The choice of alumina ceramic offers a more even distribution of pores. Pores filled with water increase the travel velocity of L-waves, but absorb T-waves. (Adapted from Roth et al. 1990)

2. Porosity of a homogeneous rock substance causes a linear decrease in wave velocity (Fig. 2.22). Velocities normally decrease with increasing porosity; however, they will increase as the water saturation of the pores increases. This relationship was established by Roth et al. (1990) using a homogeneous fine-grained ceramic with controlled porosity. Similar results were recorded on porous sandstones by Queisser et al. (1985).

3. Freeze-thaw cycles, sodium-sulfate tests, and heating-water saturation cycles were tested by Crnkovic (1982) using ultrasound after each cycle. Three rock types were examined, a gabbro and two dense limestones. The greatest loss of wave velocity occurred in the heating-water saturation test. This appears to be the test most indicative of stone durability.

Ultrasound testing appears to be a very efficient, quick, and non-destructive tool for stone evaluation. The unlimited heterogeneity of stone fabrics, however, has created many problems of interpretation. Tests should also be made of a quarry-fresh sample of the stone for comparison with the weathered stone, in order to test before it has had a chance to dilate.

## 2.12  Light Transmission

Thin, translucent slabs of stone, mostly crystalline marble and onyx-marble, have been used as nature's stained glass windows, and have been known in the sunny Mediterranean area since ancient times. The legendary "alabaster" onyx-marble windows of Galla Placidia's Mausoleum in Ravenna spread hues of soft white, orange, and pale green light. Translucent colored marble and onyx-marble panels have become fashionable again, resembling stained glass windows with nature's own marvellous geological designs (see Sect. 1.2.2.3, Fig. 1.20). At the windows of Yale University's Beinecke Rare Books Library, banded Vermont Danby marble spreads faint shades of pink to the ample interior of the building.

Light tranmission may be crudely evaluated on a qualitative basis as transparent, translucent, and opaque.

Transparent: A material is transparent if an object behind a thin slab of the material can be clearly identified, like glass, clear white mica, quartz crystals, and some gypsum crystals. Sheets of clear mica are used as windows in Indian pueblos of the southwestern USA.

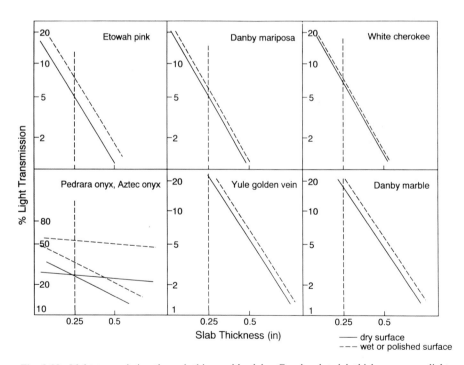

**Fig. 2.23.** Light transmission through thin marble slabs. Graphs plot slab thickness versus light transmission in percent for some translucent stone varieties. *Solid lines* Light transmission for a dry slab; *dashed lines* polished or wet surfaces. (Adapted from Winkler and Schneider 1965)

Translucent: Light is readily transmitted through thin sheets or on a thin edge, but objects cannot be identified. Many marbles and onyx-marbles are translucent.

Opaque: Light cannot pass through the material, not even in thin sheets. Most rocks are opaque.

The transmission of light through a homogeneous substance can be calculated with the equation

$$I = I_o \times r^{tx},$$

where t is the light transmittance, $I_o$ the intensity of light entering the slab, I the intensity of the transmitted light, and x the thickness of the slab (Winkler and Schneider 1965). Such measurements can readily be made with a commercial photographic spot light meter and a Kodak Neutral Grey Card for calibration of the readings. The amount of light transmitted is roughly inversely proportional to the square of the thickness of the panel.

The optical orientation of quartz, calcite, and other potentially transparent and translucent minerals influences the degree of light transmission. Light usually scatters around the crystal boundaries, where the light may either break or be dispersed by tiny gas bubbles and absorbed by pigments. A polished stone surface duplicates wetness to a certain extent. It improves the light transmission considerably, in part as a function of grain size. Light transmission is generally greater perpendicular to than parallel to the bedding planes (Fig. 2.23). Light transmission is generally very small, except for many onyx-marbles with little pigment, yet it is sufficient to be effective for special effects.

# References

Bayly B (1968) Introduction to petrology. Prentice Hall, New York, 371 pp

Brace WF (1961) Dependence of fracture strength of rocks on grain size. Proc 4th Symp Rock mechanics, Pennsylvania State Univ March 30–April 1, 1961, pp 99–103

Brace WF (1964) Brittle fracture of rocks. In: Judd WR (ed) State of stress in the earth's crust. Elsevier, pp 111–180

Choquette PW, Pray LC (1970) Nomenclature and classification of porosity in sedimentary carbonates. Am Assoc Pet Geol Bull 54(2):207–250

Colback PSB, Wiid BL (1965) The influence of moisture content on the compressive strength of rocks. Proc 3rd Can Symp on Rock Mechanics, Toronto, 1965, Mines Branch, Dept of Mines and Technical Surveys, Ottawa, pp 65–83

Conrad H, Sujata HL (1960) Dislocation theory applied to structural design problems in ceramics. Natl Acad Sci, Materials Advisory Board, Dec 1960

Crnkovic B (1982) Nondestructive method in determination of technical properties of natural stones. Durability of Building Materials 1:35–47

Cullen WC (1963) Solar heating, radiative cooling, and thermal movement–their effect on built-up roofing. Nat Bur Stand, Tech Note 231:33

Daly RA, Manger GE, Clark SP (1966) Density of rocks. In: Clark SP (ed) Handbook of physical constants. Geol Soc Am Mem 97:19–26

Deere DU (1968) Rock mechanics, geological considerations. In: Stagg KG, Zienkiewicz OC (eds) Rock mechanics in engineering practice. Wiley, London, pp 1–53

Farmer IW (1968) Engineering properties of rocks. Spon, London, 180 pp

Felix (1983) Sandstone linear swelling due to isothermal water sorption. In: Wittmann FH (ed) Werkstoffwissenschaft und Bausanierung. Lack und Chemie 1983:305–310

Fitzner B (1988) Untersuchungen der Zusammenhänge zwischen dem Hohlraumgefüge von Natursteinen und physikalischen Verwitterungsvorgängen. Mitt Ingenieurgeol Hydrogeol 29:217

Gaviglio P (1989) Longitudinal waves propagation in a limestone. The relationship between velocity and density. Rock Mech Rock Eng 22(4):299–306

Hardy HR, Jayaraman NI (1971) Hoop-stress loading – a new method of determining the tensile strength of rock. Preprints, 5th Conf Drilling and rock mechanics, Soc Petrol Eng AIME, Jan 1971, Univ Texas, SPE #3218, pp 71–83

Hawkes I, Mellor M (1970) Uniaxial testing in rock mechanics laboratories. Eng Geol 4(3):177–285

Hockman A, Kessler DW (1950) Thermal and moisture expansion studies of some domestic granites. J Res Natl Bur Stand, Res Pap RP 2087, 44:395–410

Houseknecht DW (1987) Assessing the relative importance of compaction processes and cementation to reduction of porosity in sandstones. Am Assoc Petrol Geol Bull 71(6):633–643

Hudec PP (1980) Durability of carbonate rocks as a function of their thermal expansion, water sorption, and mineralogy. ASTM STP 691. In: Sereda PJ, Litvan GG (eds) Am Soc Testing Materials, pp 497–508

Koehler W (1988) Preservation problems of Carrara marble sculptures, Potsdam-Sanssouci ("Radical structural destruction of Carrara marble"). VIth Int Congr on Deterioration and conservation of stone. Proc, Torun, 12–14 Sept, 1988, pp 653–662

Krautkraemer J, Krautkraemer H (1990) Ultrasonic testing of materials, 4th edn. Springer, Berlin Heidelberg New York, 677 pp

Mamillan M (1976) Methodes d'essais physiques pour evaluer l'alteration des pierres des monuments. In: Rossi-Manaresi R (ed) Conservation of stone, I. Bologna, June 19–21, 1975, pp 595–634

McColm IJ (1990) Ceramic hardness. Plenum Press, New York, 324 pp

McGreevey JP (1985) Thermal properties as controls on rock surface temperature maxima, and possible implications for rock weathering. Earth Surface Processes and Landforms 10:125–136

McGreevey JP, Smith RJ (1982) Salt weathering in hot deserts: observations on the design of simulation experiments. Geogr Ann 64 A:161–170

Michalopoulos AP, Triafilidis GE (1976) Influence of water on hardness, strength and compressibility of rocks. Bull Assoc Eng Geol XIII(1):1–21

Prim P, Wittmann FH (1985) Methode de mesure de l'effet consolidant de produits de traitement de la pierre. Int Félix G (ed) 5th Int Congr on Deterioration and conservation of stone. Lausanne, Sept 25–27, 1985, pp 787–794

Proctor RJ (1970) Performance of tunnel boring machines. Bull Assoc Eng Geol VI(2):105–117

Queisser A, v Platen H, Fürst M (1985) Rebound and ultrasonic on freestones of Bambeag area, FR Germany. 5th Int Congr on Deterioration and conservation of stone. Lausanne 25–27 Sept Ecole Polytechnique Federal de Lausanne, pp 79–86

Roth DJ, Stang DB, Swickard SM, DeGuire MR (1990) Review and statistical analysis of the ultrasonic velocity method for estimating the porosity fraction in polycrystalline materials. NASA Tech Memo 102501, Lewis Res Cent, Cleveland, OH 44135

Roth ES (1965) Temperature and water content as factors in desert weathering. J Geol 73(3):454–468

Skinner BJ (1966) Thermal expansion. In: Clark SP (ed) Handbook of physical constants. Geol Soc Am Mem 97:75–96

Tarkoy PJ (1981) Tunnel boring machine performance as a function of local geology. Bull Assoc Eng Geol XVIII(2):169–186

Tertsch H (1949) Festigkeitserscheinungen der Kristalle. Springer, Wien, 310 pp
Tetmajer L (1884) Methoden und Resultate der Prüfung natürlicher und künstlicher Bausteine.
    In: Mitteilungen der Anstalt für Prüfung von Baumaterialien am eidgen. Polytechnikum in
    Zürich, No 1. Commissionsverlag von Meyer und Zeller, Zürich, Switzerland, 59 pp
Volkwein A (1982) Zerstörungsfreie Prüfung von Naturwerkstein durch Ultraschall-Schwä
    chungsmessungen. Materialprüfung (Materials Testing) 24(4):119–124
Walker RD, Oence HJ, Hazlett WH, Ong WJ (1969) One cycle slow freeze test for evaluation
    of aggregate performance in frozen concrete. Natl Cooperative Highway Res Progr Rep
    65:21
Winkler EM (1986) A durability index for stone. Bull Assoc Engin Geol XXIII(3):344–347
Winkler EM, Schneider GJ (1965) Light transmission through structural marble. Am Inst
    Architects March 1965:67–68

# 3 Natural Deformation of Rock and Stone

Most rocks were exposed to stresses in the earth's crust in the geologic past. Such stresses cause cracking and faulting when brittle rocks are exposed to damaging earthquakes. Plastic folding occurs under slower long-term conditions. Stresses are compressional, tensional, or shear. The variety and combination of deformational features is complex and almost unlimited. Only the features that directly influence the stone industry will be discussed. These are esthetic appearance, safety in the quarry and on buildings, and damage to stone blocks by earthquakes.

## 3.1 Brittle Rock Fracture

### 3.1.1 Jointing

Open rock fractures are most significant. They can determine the size of the stone that can be recovered, and also the safety in the quarry. The fractures may be open, i.e., the stone separates freely, or closed if they are cemented with mineral matter introduced after cracking. Panels of decorative stone often display recemented cracks in strong color contrast with the matrix. These rock joints are the result of tension, both extension and shear. Figure 3.1 sketches the basic fracture patterns developed during compression of a cube or cylinder of stone or concrete. In quarries and rock outcrops, fracture patterns often appear almost at right angles with each other. They reflect predominant regional stress patterns (Fig. 3.2). Hodgson (1961) classifies lines of separation as systematic joints, cross joints occurring predominantly along bedding planes. Nonsystematic joints are usually curved and tend to terminate at systematic joints. Open joints determine the dimensions of sound stone (Fig. 3.2). The statistical distribution of the azimuth-oriented joints can be plotted on a compass grid like wind roses with the direction and percentage of total joints measured in the quarry. Natural stone cleavage in granites is the result of invisible preferred orientation and direction of weakness. Stone cleavage, not to be confused with mineral cleavage, facilitates the quarrying of large blocks by wedging.

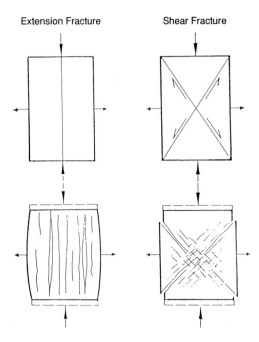

**Fig. 3.1.** Development of extension fractures and shear fractures. Extension fractures often show plumose markings on joint surfaces, whereas shear fractures may develop slickensides by minor movements

### 3.1.1.1  Fabrics on Joint Surfaces

Occasionally, systematic joints may display feather markings (Fig. 3.3) on smooth surfaces. The feathers, or plumes, diverge from a central dividing axis and continue into a plumose, feather-like system. Plumose patterns are visible on natural joint surfaces where compression or impacts have taken place (Fig. 3.5). Major plumose surfaces were observed on a large scale to both sides of a line of disturbance in a granite stone quarry (Fig. 3.4), with compression so great that channels 2 in. (5 cm) wide, cut with a kerosene-oxygen flame parallel to this line, have closed before the cutting of the block could be completed. The quarry operation had to be discontinued. Hodgson (1961) reports that splitting of slate of such joints marked with plumes initiated and continued from the chisel impact point outward through the surface as a result of tension or extension (Fig. 3.5) Plumes on slate surfaces improve the roughness for better walking safety.

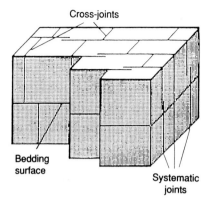

**Fig. 3.2.** Basic joint patterns. Schematic block diagram showing bedding surfaces and joint surfaces. Jointed limestone approximates the model

### 3.1.1.2 Colors on Joint Surfaces

The exposure of the open joint and crack surfaces to oxidizing or reducing water and air causes oxidation and frequent discoloring of iron and manganese with interesting displays of colors along the joint surfaces. They often filter into the adjacent stone. Some red (ferric iron) limestone-marbles have

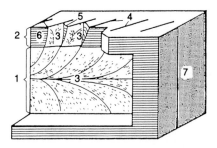

**Fig. 3.3.** Feather markings on systematic joint surfaces. *1* Main joint surface. *2* Fringe zone, prominent on large surfaces. *3* Plumose structure as barbs or feathers. *4* Border planes (joints on fringe). *5* Cross fractures, continuation of plume into fringe. *6* Shoulder of fringe to main joint face. *7* Trace of main joint face. (Hodgson 1961); modified from Roberts 1961

**Fig. 3.4.** Large feather markings on joint surface pointing away from the main fault line. Haywood granite quarry, Graniteville, Missouri

reduced the red to green ferrous iron (Fig. 3.6). Granite and limestone quarries can often deliver stone with a variety of color shades: fresh broken, seam and split face, and along joints often with a distinct brown color (Fig. 3.7).

**Fig. 3.5.** Feathers radiate from points of chisel impact in Virginia black slate. The fine relief of the feathers roughen the walking surface for better safety

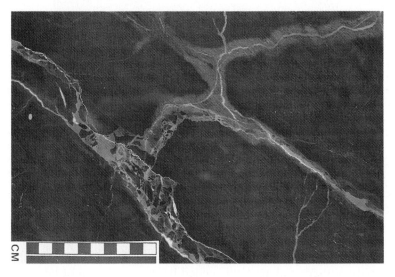

**Fig. 3.6.** Red Italian limestone-marble; cracks were later repaired with white calcite, with some brecciation. Reducing solutions have left light green margins along the cracks of ferrous iron oxide. Commercial floor tile

**Fig. 3.7.** Facing stone of storefront with different shades of color and fabrics. Rough surface is top of stylolites; tan color is from iron-stained joints. Limestone from near Chicago

### 3.1.2 Faulting

A fault is a fracture or a fracture zone along which displacement of one side relative to the other has occurred (Fig. 3.8). Displacements may range from a few millimeters to thousands of meters. Faults may have a variety of effects on stone, depending on the kind of faulting, the kind of fault filling, and the degree of rock shattering along the fault (Fig. 3.9). Soft fault gouge and rapid weathering along faults, aided by the circulation of hot gases and waters in the geologic past, should be expected. Figure 3.10 shows a boulder cut by faults and pegmatite dikes and transported from Canada to lower latitudes by the ice masses thousands of years ago. Such glacial erratics are frequently found and used as accents for landscaping in many former

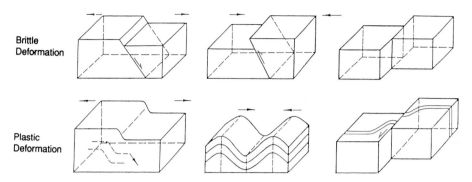

**Fig. 3.8.** Principles of rock deformation found in stone faulting and folding. Brittle deformation: tension leads to normal faulting, compression to reverse faulting, and shear to strike-slip faulting. Plastic deformation: tension leads to flexures and monoclines, compression to folding (both symmetric and asymmetric), and shear to drag folding

**Fig. 3.9.** Limestone cracked along the crest of fold axis, at about 45° of fold axis. Cracks were cemented with white calcite. Bottom of crushed stone quarry, near Romney, West Virginia

**Fig. 3.10.** Faulting of boulder in granite, banded gneiss. Decorative boulder for landscaping; normal and reverse faulting along pegmatite dike. Boulders of this type are often seen in commercial landscaping. Notre Dame University campus, Indiana

glaciated areas of the world. Brecciation caused by crushing along faults is commonly seen on some limestone-marble veneer surfaces; slabs often break along such zones during cutting and milling if these have failed to recement properly in nature (Fig. 3.6). Polished and slickensided, or streaked, surfaces may replace brecciation, often with secondary white calcite filling along a fault. The direction of the relative movement can be determined by measuring slickensides in the outcrop. Fault filling with white calcite or quartz is common and decorative when polished (Figs. 3.11 & 3.12). Naturally cemented fault zones occur where crushing was minor.

## 3.2  Active Rock Pressure

Minor rock deformations in shallow quarries often take place as rock bursts or as slow creep. Rock in its natural setting establishes a temporary equilibrium with the external forces of erosion and weathering. An open pit quarry defies man's interference, often with violent reactions. It tends to reestablish equilibrium by the redistribution and reorganization of stresses.

**Fig. 3.11.** Two generations of small faults in a gray limestone-marble; an earlier generation with white calcite fillings, and a later fault filled with black carbon residue, offsetting the earlier faults. Intensive fracturing reflects the mountain-building process of the Atlas mountains, Morocco. Paneling of Memorial Library, Notre Dame, Indiana

The response may take hundreds of years – or only minutes. For example, vertical cut by saw or by torch may close while cutting is still in progress.

Rock pressures in quarries are generally ascribed to mainly two sources, stress relief by eroded overburden and true mountain-building or tectonic pressure.

### 3.2.1 Stress Relief

Stress in rock is in equilibrium when neither normal nor shear stresses are transmitted through its surface (Voight 1966). Internal stresses develop in a

**Fig. 3.12.** Densely fractured and sheared white Alaska marble; fractures are filled with black graphite. Old store-front, Colorado Springs, Colorado

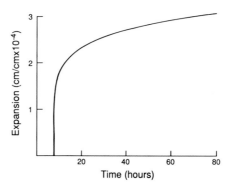

**Fig. 3.13.** Volume expansion of marble – or granite – with time. (After Voight 1966)

rock body through mountain-building events. The removal of confining stress during quarry operations gradually leads to expansion toward the original prestressed condition, like relaxing a squeezed spring. The expansion of prestressed granites, crystalline marble, or gneisses may take years to recover. The rate can be fast at first, but then diminishes with time.

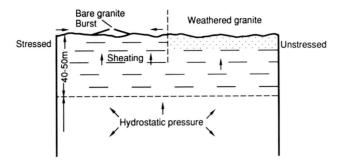

**Fig. 3.14.** Stress distribution and stress relief near the terrain surface in unweathered and weathered granite. (After Kieslinger 1960)

**Fig. 3.15.** Sheeting of granite. Separation of sheets parallel to the land surface; layers are thinner toward the land surface. Marble Falls, Texas

Voight's data were collected during 80 h of expansion (Fig. 3.13). A theoretical stress distribution in a granite quarry is sketched after Kieslinger (1960) (Fig. 3.14). Residual stresses affecting bare rock surfaces can develop sheeting (exfoliations) and bursting by horizontal compressional forces (Figs. 3.14–3.16). Stress relief developed along distinct horizontal joints which lie approximately parallel to the original land surface as unloading of the overburden by erosion began some time in the geologic past. Sheeting

**Fig. 3.16.** Stress relief on a bare rock surface by slabbing or tenting, also known as rock bursts or bumps. White granite from this quarry replaces bowing Carrara marble on 93 floors of the Amoco building in Chicago (see Fig. 7.23). Will the granite from this quarry resist bowing? Mt. Airy quarry, Mt. Airy, North Carolina

intensifies toward the land surface to form major exfoliation surfaces and exfoliation domes (Fig. 3.15). Thin sheets of granite separate on the bare rock floor of the Mt. Airy granite quarry near Winston-Salem, North Carolina (Fig. 3.16). The process is also called "tenting". In a similar way, exfoliation is frequently observed on granite base courses in buildings near the ground level, accelerated by rising ground moisture and especially in the presence of salts from the soil or deicing compounds (Fig. 3.17); it also occurs where stone is continuously wetted, e.g., on stone fountains (Fig. 3.18). Stress relief is also frequently observed in geologically prestressed stone slabs mounted as curtain walls. Unstressing may be a relatively quick process in humid climates, often occurring in less than 15 years of exposure. The bowing of well-anchored vertical marble panels is complex as it involves several processes, starting with acid-rain surface attack. Granite panels behave in a similar way, except for the origin of the pore system by contraction when cooled (see Chap. 10) and the lack of attack by acid.

———————————————————————————————▶

**Fig. 3.18.** Fountain in a boulder of granite: spalling by action of water triggered by locked-in stresses. Bamberg, Germany

**Fig. 3.17.** Scaling of granite by moisture; salt action triggered locked-in stresses. Granite from Stone Mountain stock, Georgia. Stone Mountain is known for sheeting of granite; the stone contractor should have received a warning. Base course of the Chicago Municipal Library, Chicago, Illinois

**Fig. 3.19.** Vertical cracks in ionic-type ribs, opened up by the relief of locked-in stresses (see also Fig. 3.20). Multiple cracking dissociates the calcite grains from the rock, Georgia marble. The photo covers the entire width of the rib (35 mm). South side of the Field Museum of Natural History, Chicago, Illinois

Rapid stress relief can lead to micro-cracking and unfavorable changes of some of the physical properties (Obert 1962). Winter freezing and heating can accelerate the process of unstressing. Mt. Airy granite has been used to replace warped white Carrara marble panels on Chicago's Amoco building after only 15 years of exposure to the weather (Figs. 3.16, 7.20 & 7.21; see also Sect. 7.3). The granite will still warp like the marble; however, it will take more time.

Sheeting tends to ease quarry operations, but it also permits the entry of the important weathering agents, air and moisture. Blocks of dimension stone should be stored outside to "age" the rock. This "weathering" for months before actual use serves to relieve stress and cure the stone through case hardening (Sect. 6.6). The full evaluation of stress conditions, including the directions of major stresses, should be performed at all quarry sites. The degree of stress relief may vary in different regions, and even in different parts of the same quarry.

Stress relief may also be observed on ribs and edges where expansion in prestressed crystalline marble and granite is possible. The load of supporting columns of a building can overlap with true stress relief. Encasement of Chicago's Field Museum of Natural History building with scaffolding for masonry cleaning permitted such a study of the white Georgia marble Ionic

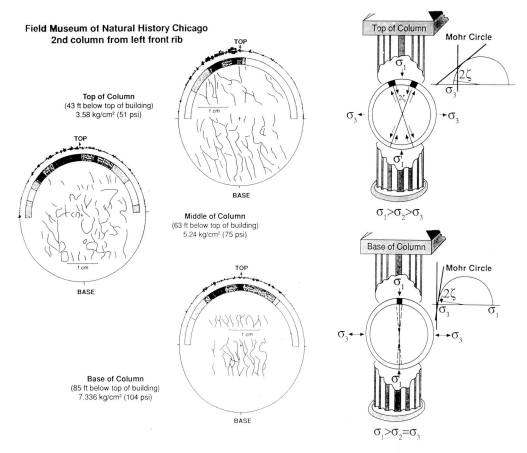

**Fig. 3.20.** Plot of cracks on flat-topped ribs in coarse-grained Georgia marble. The cracks were plotted with reference to the vertical axis of the columns for three different heights; the patterns of the principal stresses are plotted for the top and bottom readings of the columns, with the resulting Mohr circles. South side of Field Museum of Natural History, Chicago, Illinois. (Winkler and Kirchmayer 1989)

columns. Parts of these columns were exposed to acid rain for 70 years (see Sect. 7.8; Winkler and Kirchmayer 1989). The orientation of open cracks relative to the vertical column axis was plotted from the flat rib-tops using vertical stereo-macro-photographs. Such a plot reflects the interaction of the rock's present stresses and those attributed to the current load of the building resting on the columns, with considerations of the original rock structure, e.g., foliation, which can influence the angle of the direction of cracks (Fig. 3.19) with the vertical column axis. Figure 3.20 plots the shear forces at three height levels based on the direction of cracks from the stereo-photographs. Mohr circles are constructed to show the difference in principal stresses between the top and the bottom of the sample columns. The study indicates that only 7–12% of the stresses are inherited from the geological past. The remaining stress is the load of the building itself.

Cutting of stone slabs from large blocks releases most of the stresses. But stresses may also be added to a stone by the processes of grinding, polishing, and flame finishing. Modern polishing machines reduce such stressing.


## 3.3 Plastic Deformation; Creep

Plastic deformation in rock takes place when tension, compression, or shear stresses act slowly, so that the rock does not break but flows under the constant stress. Pressure and heat facilitate this process. Creep is accelerated by well-developed perfect cleavage of the mineral components, e.g., under their own weight calcite in crystalline marble will show creep when suspended as stone slabs (cladding). Plastic deformation can also occur at surface temperatures and pressure by the load of the stone slab (Fig. 3.21). Creep rates are controlled by the modulus of elasticity of the rock substance (Table 3.1). Flat-lying slabs or bars of marble behave in similar ways. Yet, it is difficult to determine whether such a deformation is from stress relief,

**Fig. 3.21.** Plastic deformation in dolomite-rock growing into the tunnel opening. Entrance to quarry in South St. Louis, Missouri

**Table 3.1.** E values and creep rates at various states of elasticity (After Farmer 1968)

| Rock type | E value $(kg/cm^2 \times 10^5)$ | Creep rate $(100\,kg/cm^2)/10$ years |
|---|---|---|
| Quasielastic basalt, siltstone | 12 | $7.6 \times 10^{-7}$ |
|  | 10 | $1.0 \times 10^{-6}$ |
|  | 8 | $1.4 \times 10^{-6}$ |
| Semielastic marble, sandstone | 6 | $2.1 \times 10^{-6}$ |
|  | 4 | $4.0 \times 10^{-6}$ |
| Nonelastic schist, rock salt | 2 | $1.1 \times 10^{-5}$ |
|  | 0.5 | $8.9 \times 10^{-5}$ |

heat–cold dilation, or true plastic deformation by creep, or a combination of all these factors. Long-range creep tests in the laboratory over 10 years show creep is an undulating process, usually interpreted as initial elasticity for 200 to 300 days, with the amount of deformation as 10–45% of the intial elastic deformation (Ito and Sasajima 1987).

Warping of vertically mounted thin marble panels is not caused by true plastic deformation, although some creep may be involved. Normally, dilation by temperature changes and hygric forces are the main causes (Sect. 7.3).

**Fig. 3.22.** Intensely folded crystalline marble, Italian Rosso Luana, used as a decorative tabletop

### 3.3.1 Folding

Slow deformation under plastic conditions may lead to folding. This is found in all rock types. Plastic conditions prevail in depth when compressional, tensional, and shear folding occur on both large and small scales (Fig. 3.8). Combinations of faulting and folding are common. Folding can be frequently seen in quarry walls, and even on a smaller scale in marble tabletops (Fig. 3.22). Flowfolding can form complex patterns on a small surface. Figure 3.22 shows a tightly and complexly flowfolded crystalline marble in a decorative tabletop. Spectacular flow patterns can be found along the margin of granitic bodies (see Figs. 1.5 & 1.9).

## 3.4 Damage to Stone by Blasting and Earthquake Shocks

Much damage has been inflicted to natural stone on buildings by earthquake shocks but also by unskillful blasting and pile driving. The magnitude of the explosive charge and the direction of the shock determine the kind and degree of damage.

### 3.4.1 Damage to Stone by Blasting

#### 3.4.1.1 Shock Action

Stone fragmentation and microcracking by blasting can cause macroscopic and microscopic damage. Quarry blasting is intended to aid the separation of large blocks where natural lines of separation, like jointing and sheeting, are missing. Upon detonation, the explosive charge sends out a stress pulse; stone absorbs and stores the excess energy. The reflection of this stress pulse at a free face, e.g., bedding planes or joints, and the expansion of the explosive gases combine to result in fragmentation. Fractures from un-loading can develop as the pressure of the blast declines. The stored elastic potential energy is subsequently released. The release of this residual stress and the development of fractures depends on the quantity of stored energy and on the brittleness of the rock. Mueller (1964) sketches shock-wave encounters parallel to and at 45° to fractures (Fig. 3.23). If the stone is moved quickly from the production line to its final placement, the integrity of the stone can deteriorate, fractures can develop, and permeability increases, allowing further damage to develop rapidly (Lienhart and Stransky 1981). Lienhart approximates the energy imparted to the rock during loading with the equation:

$$P = 4.18 \times 10^{-7} \, (DC^2/1 + 0.8\,D),$$

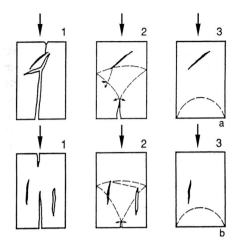

**Fig. 3.23.** Direction of shockwaves. Opening of fractures by an advancing blast wave. **a** Shock-wave encounter runs parallel to fracture; **b** shock-wave encounter runs at an angle of near 45°. (Mueller 1964)

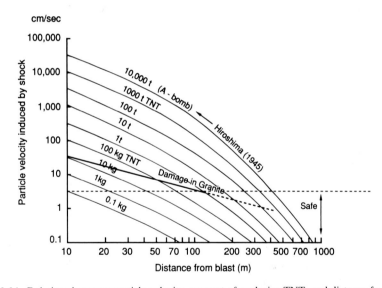

**Fig. 3.24.** Relations between particle velocity, amount of explosive TNT, and distance from the blast. Approximate line of structural safety is marked by a *dashed line* at 5 cm/s blast-induced particle velocity. The *full line* indicates where granite is damaged. (Adapted after Langer 1965)

where P is detonation pressure in bars (1 bar = 14.5 psi), D is specific gravity of the explosive, and C is detonation velocity (ft/s).

The sonic velocity of some stones, e.g., granite or basalt, should be matched with the sonic detonation velocity of the explosive in order to create the most effective fragmentation. Figure 3.24 gives the ranges of common rock types. The relationship between the particle velocity, the distance from the shot (blast) point, and the magnitude of the TNT equivalents can be scaled from Langer's (1965) nomogram in Fig. 3.24. This graph shows shock safety in mines and quarries with the maximum range of the effect of microfracturing in stone, 135 m for a 500-kg TNT explosive charge, and more than 350 m maximum distance for a 10-kt charge (Short 1961). A particle acceleration of 5 cm/s appears to be a safe shock velocity for most rocks used in the stone industry. Dimension stone quarries avoid blasting to prevent unnecessary shattering and microfracturing. Blasting with minimum vibration is used only to produce crushed stone. Shocks may also be generated by exploding bombs. While no attention is paid in this book to general structural damage, fracturing of stone can be complex. Detailed studies of rock blasting in quarries and tunnels have been performed at the US Bureau of Mines by Duvall and Atchison (1957), and by Siskind et al. (1980). The shock wave advances from the shot point with compressive stresses till the waves reflect from free surfaces as tensile stresses toward the shot point. This can cause "slabbing" at bedding planes and joint surfaces. Rock failure can occur above 2 cm/s at very low frequency (below 40 cm/s). The tensile strength of stone amounts to only a fraction of the compressive strength. The sequence of rock fissuring is also illustrated in Fig. 3.23.

Since no explosives should be used in dimension stone quarries, blasting with minimum vibration is used only for crushed stone. Line drilling involving serial perforations with 2- to 3-in.-diameter bore holes spaced two to four times their diameter for maximum efficiency will make blasting unnecessary.

### 3.4.2 Vibration by Earthquakes and Other Sources

Strong vibrations by heavy truck traffic, pile driving, or explosions may be as damaging to buildings and building stones as earthquakes. Rotation and crushing of dimension stone blocks can be observed on columns and buildings (Figs. 3.25 & 3.26). Voltaire pictures total destruction following the Lisbon 1755 earthquake (Fig. 3.27). The vibration by shock is expressed by the peak particle velocity (ppv, in cm/s), which decreases with the distance from the source, and by acceleration due to the force of gravity ($g$) ($1g = 980 \text{ cm/s}^2$), while the ppv is used as the indicator for the damage potential to a structure, with a greater velocity in hard rock than in soft soils. The acceleration at low frequencies generates forces between the soil and the foundations of buildings, mainly due to soil compression or liquification. The acceleration is about equal between rock and most soils, but tends to amplify by a factor of 1.5 on marshy ground. The direction of acceleration depends on the characteristic of the shocks. Seismographs, which record shock waves, are set up so that all directions of incoming waves can be recorded. Figure 3.28 sketches the principle of such a recording configuration. Three types of waves with different times of travel are emitted from

**Fig. 3.25.** Cracking of stone blocks by an earthquake. Vertical acceleration affected lower courses as the mortar decreased the shocks upward. Cathedral of Kalamata, Greece. Earthquake in 1986 measured M = 6.0; only 20 people died; MM = VIII

blasts and earthquakes: The primary or compressional waves (P), which travel up to 7.5 km/s; shear or transverse waves (S), which travel about half the speed of the P waves; and long or surface waves (L), which are slowest. Ultrasound-emitting instruments (20 kHz to 1.5 MHz), often used as flaw detectors, emit and receive similar waves. The travel velocity and absorption of such waves give insight into the density of the material and the possible presence of cracks. Two different modes of presentation characterize earthquake intensity; the Richter magnitude scale (M), measuring the release of the total energy at the source of the shock, and the modified Mercalli scale (MM), measuring the degree of damage. While the Richter scale refers to the total energy release at the source, expressed as a number from 1 to 9, the MM scale records the degree of damage to buildings or monuments at the place of damage.

### 3.4.2.1 Richter Magnitude (M)

The Richter scale compares earthquakes throughout the world based on the energy release, the Richter magnitude, M. The basic scale consists of the log A (wave amplitude) minus the log $A_o$ distance factor from the epicenter to

**Fig. 3.26.** Section of ancient column, rotated and dislocated by many earthquakes in 2400 years of weathering exposure. Acropolis, Athens, Greece

the recording station (Fig. 3.29). The total energy emitted increases 32 times for each unit of M; M = 8.9 appears to be the upper limit of the strain energy stored in rock as rupture occurs.

In contrast, the M value usually does not correspond with the degree of damage at a given location. Table 3.2 lists a few important historical earthquakes.

Casualty and M figures are approximations. The M value is not in proportion to the casualties and to the degree of damage, which depends on the depth of the source, the composition of the rock or soil (soft soil or hard rock), and the presence of faults.

**Fig. 3.27.** Earthquake in Lisbon, Portugal (1755), as pictured by Voltaire in Cancice – a pessimist's view. (US Geological Survey 1972)

### 3.4.2.2 Modified Mercalli (MM) Intensity Scale

The original Mercalli scale, devised in 1902, was later modified (1931) to the MM (modified Mercalli) scale. The intensity is a measure of the damage at the place of observation, regardless of the distance from the epicenter of the shock. Krinitzsky and Chang (1988) relate the scale directly to the needs of the engineer. Table 3.3 gives the 12 stages of the MM scale, and the net effect of the earthquake for each step.

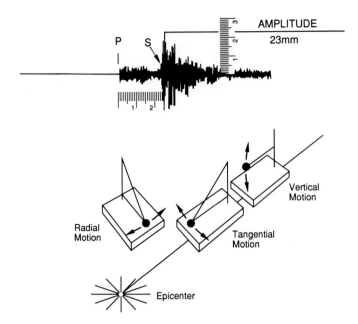

**Fig. 3.28.** The Richter scale. Seismogram of a hypothetical earthquake and the directional orientation of a seismograph. Timing and wave amplitude can be scaled off the graph, as can the onset of the primary or compressional (*P*) waves, the shear or secondary (*S*) waves, and the longitudinal (*L*) or surface waves (not marked here). (Sherburne 1977)

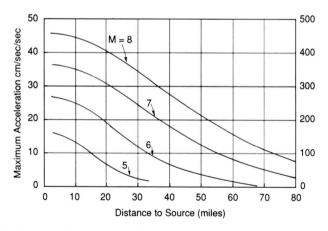

**Fig. 3.29.** Maximum acceleration of earthquake waves as a function of the distance from the source of the shock, for each magnitude M

**Table 3.2.** Historical earthquakes

| Location | Year | M | MM | Casualties | Damage |
|---|---|---|---|---|---|
| China | 1556 | 7 (?) | XII | 850 000 | Breaking of dikes |
| Lisbon (Portugal) (Fig. 3.27) | 1755 | 9 (?) | XII | 60 000 | City destroyed |
| San Francisco (USA) | 1906 | 8.2 | X | 700 | Fires, city destroyed |
| Tokyo (Japan) | 1923 | 8.3 | XI | 143 000 | City destroyed |
| Anchorage (Alaska) | 1964 | 8.6 | X | 102 | Heavy damage |
| Mexico City (Mexico) | 1985 | 8.1 | IX | 9 500 | Structures on soft ground |
| Kalamata (Greece) | 1986 | 6.0 | VIII | 10 000 | Heavy damage |
| Armenia (former USSR) | 1988 | 6.8 | IX | 25 000 | Total destruction |
| Loma Prieta (USA) | 1989 | 7.1 | | 56 | Much damage |
| Western Iran | 1990 | 7.7 | XI | 40 000–50 000 | Total destruction |

Data from file of the National Earthquake Information Center, Boulder, Colorado.

**Table 3.3.** The stages of the MM scale

    I. Not felt
   II. Felt by persons at rest on upper floors
  III. Vibration like the passing of a light truck; hanging objects swing
  IV. Vibration like the passing of a heavy truck; sensation of a jolt, like a heavy ball striking the wall; windows, dishes, and doors rattle
   V. Felt outdoors; sleepers wakened; liquids spill; small unstable objects are displaced or upset; doors swing
  VI. Felt by all, many frightened; dishes break; objects fall off shelves; furniture moves or is overturned. Weak plaster or masonry D cracked
 VII. Difficult to stand; drivers in motor cars notice; furniture breakage. Damage to masonry D, including cracks
VIII. Damage to masonry C, partial collapse; some damage to masonry B, but none to masonry A; fall of stucco and masonry walls; twisting of columns, chimneys, and statues
  IX. General panic; masonry D destroyed; masonry C heavily damaged, often complete collapse. Frame structures shifted off foundations, if not bolted; serious damage to reservoirs, cracks in ground
   X. Most masonry and frame structures destroyed; some well-built wooden structures and bridges destroyed; rails bent
  XI. Rails and pipelines break
 XII. Destruction total; large landslides

Quality masonry Classification

Masonry A: Good workmanship, mortar, and design; reinforced laterally, designed to resist lateral forces

Masonry B: Good workmanship and mortar; reinforced but not designed to resist lateral forces

Masonry C: Ordinary workmanship and mortar; no extreme weakness, but neither reinforced nor designed to resist lateral forces

**Table 3.4.** Correlation of the Richter (M) with the Mercalli (MM) scale (Krinitzsky and Chang 1988)

| Magnitude (M) | MM Max. intensity | Radius of near field (km) |
| --- | --- | --- |
| 5.0 | VI | 5 |
| 5.5 | VII | 15 |
| 6.0 | VIII | 25 |
| 6.5 | IX | 35 |
| 7.0 | X | 40 |
| 7.5 | XI | 45 |

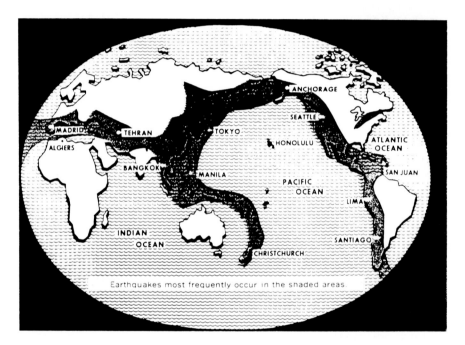

**Fig. 3.30.** Earthquake-prone areas of the world. St. Lawrence Seaway into Tennessee was added by the author. (US Geological Survey 1972)

Masonry D: Weak materials, such as adobe; poor mortar; poor workmanship

A correlation of M with the approximate MM values is possible for sites near the source (Table 3.4). Only the Richter scale corrects with its distance factor for sites far away.

The vibrations are usually given as peak particle velocities (ppv), which is the maximum velocity attained by a given wave; this is a leading indicator of the potential damage to a structure. Sedovic (1984) sets safety levels for historic structures in terms of ppv as follows:

0.5 in./s (1.25 cm/s): safe level of vibration for historic structures.

0.5–0.75 in./s (1.25–1.88 cm/s): unacceptable for historic structures.

2.0 in./s (5 cm/s): unacceptable for historic buildings, and safety limit for modern buildings.

The value 5 cm/s corresponds approximately to MM = IV. Figure 3.29 gives the approximate ppv values for the different MM numbers (Krinitzsky and Chang 1988). Figure 3.26 illustrates earthquake-induced rotation of blocks in marble columns at the Parthenon in Athens. Similar rotation was also observed on the temple of Apollo Epikourios (Fig. 7.17). Vertical acceleration in the well-built cathedral of Kalamata, Greece, has visibly cracked the masonry and spalled the blocks of the lower courses, while the upper courses remained intact where the shock waves were evidently cushioned by the mortar joints (Fig. 3.25).

Many historic monuments are sited in earthquake-prone areas of the world. The restoration architect should be aware of these regions (Fig. 3.30). Local state and provincial agencies usually have information on previous earthquakes.

# References

Duvall WI, Atchison C (1957) Rock breakage by explosives. US Bur Mines, Rep Invest #5356:52

Farmer IW (1968) Engineering properties of rocks. Spoon, London, 180 pp

Hodgson RA (1961) Classification of structures on joint surfaces. Am J Sc 259:493–502

Ito H, Sasajima S (1987) A ten year creep experiment on small rock specimens. Int J Rock Mech Miner Sci 24(2):113–121

Kieslinger A (1960) Gesteinsspannungen und ihre technischen Auswirkungen. Z Dtsch Geol Ges 112(1):164–170

Kieslinger A (1967) Residual stress, summary. Proc 1st Contr Int Rock Mech III:354–357

Krinitzsky EL, Chang FK (1988) Intensity-related earthquake ground motion. Bull Assoc Eng Geol XXV(4):425–439

Langer M (1965) Die Dämpfung von Druckwellen im Gebirgskörper. Z Zivilschutz, Baulicher Zivilschutz 29(5):2–6

Lienhart DA, Stransky TE (1981) Evaluation of potential sources of riprap and armor stone – methods and considerations. Assoc Eng Geol Bull XVIII(3):323–332

Mueller L (1964) Beeinflussung der Gebirgsfestigkeit durch Sprengarbeiten. Rock Mech Eng Geol Suppl I:162–177

Obert L (1962) Effects of stress relief and other changes in stress on the physical properties of rock. US Bur Mines Rep Invest 6053:5

Roberts JC (1961) Feather fracture and the mechanics of rock-jointing. Am J Sci 259:481–492

Sedovic W (1984) Assessing the effect of vibration on historic buildings. Bull Assoc Preserv Tech XVI(3, 4):53–61

Sherburne RW (1977) Earthquake magnitude determination. Calif Geol 30:161

Short NM (1961) Fracturing of rock salt by a contained high explosive. Drilling and Blasting Symposium. Q Colorado School of Mines 56(1):221–257

Siskind DE, Stagg MS, Kopp JW, Dowding CH (1980) Structure response and damage produced by ground vibration from surface mine blasting. US Dep Inter Bur Mines Rep Invest 8507:74

US Geological Survey (1972) Earthquakes. US Dep Inter Geol Surv USGS: Inf-69-4 (R-5)

Voight B (1966) Residual stresses in rocks. Proc 1st Congr Int Soc Rock Mech, Laboratorio Nacioual de Eugenlearia Civil, pp 45–50

Winkler EM (1984) Buttressed expansion of granite and development of grus in central Texas: discussion. Z Geomorphol NF 28(3):383–384

Winkler EM (1988) Weathering of crystalline marble. In: Marinos PG, Koukis GC (eds) Engineering geology of ancient works, and historical sites. Balkema, Rotterdam, pp 717–721

Winkler EM, Kirchmayer M (1989) Stress analyses from stereo-P, Field Museum Natural History, Chicago. Geol Soc Am 21(4):17

# 4 Color and Color Stability of Stone

The color of structural and monumental stone has challenged the architect for the most effective and harmonious appearance in architectural design since ancient times. The utilization of different color shades of stone has given new life to many existing structures. Stone colors are influenced by the color of the predominant mineral, but also by the adjacent minerals, grain size, and grain cement.

All rocks have their own characteristics: there are the stable pigments of the igneous rocks, e.g., granites; the variable, warm and frequently unstable pigments of the sedimentary rocks, e.g., sandstones, limestones, and limestone-marbles; and the quite different cold colors of the true metamorphic rocks, e.g., slates, marbles, and serpentines. Each rock group is characterized by a range of colors and textures. The architect should therefore be aware of the different possibilities of color variations and pigment stabilities. Changing colors can occur quite fast in sedimentary rocks. Polishing a stone surface or wetting, however, does not change the color hue, just its saturation.

## 4.1 Presentation of Color

A simple and unified presentation of color is of great importance to the color-conscious architect for purposes of accurate color description and comparison, matching for stone replacement, recording of possible change by exposure to the atmosphere or rainwater, etc. Several systems of color comparison developed by physicists are summarized by Wright (1969). The Munsell color system, however, is the most practical method for characterizing color for building and decorative stone, and is generally accepted today. The Munsell system is here discussed as the simplest to understand and use, without the use of complex and expensive colorimeters.

The Munsell color system, first devised in 1915, arranges the three attributes of color, namely, hue, chroma, and value, into orderly scales of equal visual steps by which color may be analyzed and described accurately as under standard light conditions (Figs. 4.1 & 4.2). Under these conditions the hue, chroma, and value of the color chips correlate closely with those of the actual stone. Escadafal et al. (1989) compare and correlate both

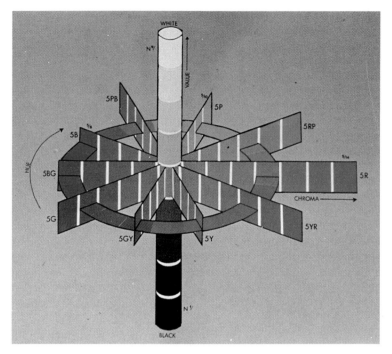

**Fig. 4.1.** Munsell color wheel: Hue, value, and chroma in relation to color space. The values of white to black, and the chroma from gray to red are used as an example of the Munsell color presentation. (*Munsell Book of Colors* 1947)

techniques in the description of soil colors based on organic and iron contents.

### 4.1.1 Hue

The chromatic colors in the Munsell system are divided into five principal classes: red (R), yellow (Y), green (G), blue (B), and purple (P). A further division yields five intermediate stages: yellow–red (YR), green–yellow (GY), blue–green (BG), purple–blue (PB), and red–purple (RP). These are combinations of the five principal hues. The hues extend around a horizontal color sphere about a neutral or achromatic vertical axis. When finer subdivisions are needed the ten hue names and symbols may again be combined as red–yellow–red (R-YR). For finer divisions, the hues may be divided into ten steps, 1R to 10R. The designation 5R marks the middle of the red hue, 1R the faintest, almost gray, and 10R the strongest, deepest red. Further refinements are given in Table 4.1. Figures 4.1 and 4.2 give the position of the three attributes. Nimeroff (1968) attempts a numeric quantification of the hues plotted on a circle with 100 subdivisions (Fig. 4.3).

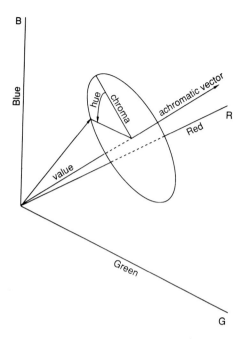

**Fig. 4.2.** Schematic representation of the Munsell coordinates *R*, *G*, and *B* (red, green, and blue). (After Escadafal et al. 1989)

**Table 4.1.** Hue names used in the Munsell color chart

| Name | Abbreviation | Name | Abbreviation |
|------|-------------|------|-------------|
| Red | R | Purple | P |
| Reddish orange | rO | Reddish purple | rP |
| Orange | O | Purplish red | pR |
| Orange yellow | OY | Purplish pink | pPK |
| Yellow | Y | Pink | PK |
| Greenish yellow | gY | Yellowish pink | yPK |
| Yellow green | YG | Brownish pink | brPK |
| Yellowish green | yG | Brownish orange | brO |
| Green | G | Reddish brown | rBr |
| Bluish green | bG | Brown | Br |
| Greenish blue | gB | Yellowish brown | yBr |
| Blue | B | Olive brown | olBr |
| Purplish blue | pB | Olive | Ol |
| Violet | V | Olive green | OlG |

From Nickerson Color Fan, 40 hues maximum chroma, Munsell Color Company, Inc. Baltimore 2, Maryland.

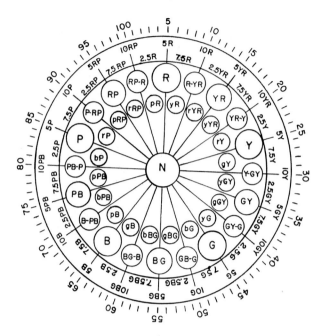

**Fig. 4.3.** Quantification of the Munsell hue notation by the subdivision of the full circle into 100 parts. (Nimeroff 1968)

### 4.1.2 Chroma

The chroma of a color indicates the strength (saturation) or the degree of departure of a particular hue from neutral gray of the same value. The scales of chroma extend from /0 for a neutral gray to /10, /12, and /14. Figure 4.4 presents the purple section of the Munsell color solid.

### 4.1.3 Value

The value notation indicates the degree of lightness or darkness of a color in relation to a neutral gray scale, which extends in a vertical direction from a theoretically pure black, symbolized as 0/, at the bottom to a pure white, symbolized as 10/, at the top. A gray of achromatic color that appears visually halfway in lightness between pure black and pure white has a value notation of 5/. Lighter colors are indicated by numbers above 5, while darker ones are indicated with numbers below 5 (Fig. 4.1). Table 4.2 correlates the ten Munsell values with the commercial Kodak Gray Scale (part of a color separation guide for photographic purposes) and with the true light reflectance measured as a percentage of daylight. This correlation quantifies the Munsell value scale.

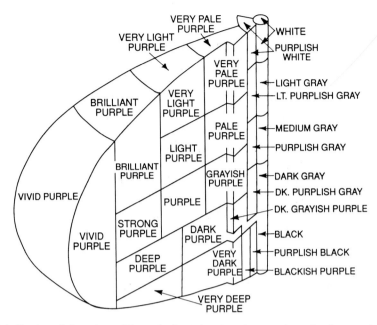

**Fig. 4.4.** Portion of the color solid purple from the neutral gray axis to the deepest chroma. (Kelly and Judd 1976)

**Table 4.2.** Quantitative comparison of the Munsell lightness (value) with the Kodak Gray Step Scale, and the absolute light reflectance

| Value | Munsell value | Kodak Gray Step Scale | Reflectance | |
|---|---|---|---|---|
| | | | MgO | Absolute |
| White | 9 | 0.10 | 78.66 | 77.4 |
| | 8 | 0.20 | 59.10 | 59.6 |
| | 7 | 0.30 | 43.06 | 43.1 |
| | 6 | 0.50 | 30.05 | 30.7 |
| | 5 | 0.70 | 19.77 | 20.1 |
| | 4 | 1.00 | 12.00 | 11.8 |
| | 3 | 1.30 | 6.55 | 6.5 |
| | 2 | 1.60 | 3.13 | 2.8 |
| Kodak Neutral Gray Card, gray side | | | 19.8 | 19.2 |
| | | white side | 85.0 | 85.6 |
| Black | 1 | 1.90 | 1.21 | |

From Munsell (1976), and Kodak Gray Step Scale (1963)

The color medium red serves as an example for the three color attributes explained by Kelly and Judd (1976) (Figs. 4.1, 4.2 & 4.4). The complete Munsell notation for any chromatic color is written as hue, value/chroma. For example, *5R 5.5/6.5R* is located in the middle of the red hue; 5.5/ is the lightness of Munsell value near the middle of light and dark; 6 is the degree of Munsell chroma or color saturation, which is in the middle of saturation. Decimals may be used, such as 2.5R 4.5/2.4, whenever a finer division is needed for any of the three attributes.

The Geological Society of America has selected the *Munsell Book of Colors* as the basis for the simple *Rock Color Chart* (Goddard et al. 1948). The chart is a semiquantitative tool for the field and laboratory; it is a small-format 6-page folder with 115 color chips measuring 0.5 by 0.75 in. Each sheet combines two or more adjacent hues. Experience helps when using such a color identification chart in the field. Based on the same principle and format, the US Department of Agriculture has accepted a more complete

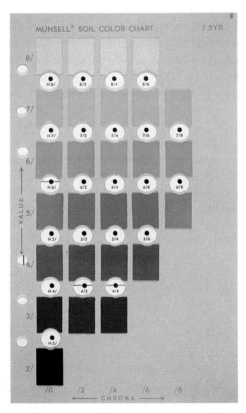

**Fig. 4.5.** A page of the *Soil Color Chart*. The *round holes* below each color chip are used to view and compare the stone, rock, or soil sample behind the chart. Value and saturation are written into the holes. (Munsell 1973)

version as the *Soil Color Chart* with 196 color chips of the same size. Each sheet is restricted to a single hue. The greater volume of the chart is originally meant to be used to describe soil colors. The US Department of Agriculture has included the *Soil Color Chart* into the *Soil Survey Manual* (Fig. 4.5). To take color readings with both color charts it is possible to isolate the color for comparison while the other color chips are covered. Coarse-grained sedimentary rocks, like conglomerates, or coarse-grained granites should be viewed from a distance in order to get a monotone that can be matched against the chips on the rock or soil color charts. Judd and Wyszeski (1963), and Kelly and Judd (1976) provide more technical information on color. The data were readily obtained with simple instruments, as follows.

### 4.1.4 Hue, Chroma

It is easy to evaluate hue, chroma, and saturation by visual comparison with the help of the *Rock Color Chart* (Fig. 4.5). Nimeroff (1968) quantified the hues by placing the hues on a graduated circle as the groundplan of the color wheel, starting with 5R at 5, 10-R at 10, 10Y at 30, etc. to 100. Figure 4.3 shows Nimeroff's scheme for practical use.

### 4.1.5 Value

The lightness, reflectance, or value appears to be the only attribute of color which can be measured accurately with a good photographic light meter. The determination of the lightness of a colored rock surface excludes the color perception. Winkler (1979) uses the gray side of the Kodak Grey Card as a reference reading of near 20% light reflection on a photographic light meter. The lens opening or diaphragm is set at F-4.0 or 5.6, marked in Fig. 4.6. The exposure meter is first directed against the gray side of the 8 × 10 in. Kodak card. The light meter should have at least a 10°, and preferably a 1°, reading of the viewing area. The adjustment to the zero position of the needle for the ambient light reflection is performed with both the shutter speed and the ASA setting, whereby the test card is to be set up side-by-side at the same angle to the sun as the object. The light meter direction is shifted from the test card to the object and the new reading taken and plotted on the abscissa of the chart; the reflection is read on the ordinate axis. More details of this technique may be obtained from Winkler (1979). The darkening of a stone surface by weathering, stone consolidation, or waterproofing can be recorded as light reflection with readings obtained before and after treatment. Greater accuracy may be obtained with separate spot meters, which can record an angle of 1° view, and readings in absolute

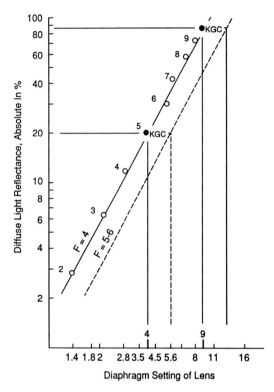

**Fig. 4.6.** Reflectance versus the diaphragm opening of the photographic camera. Ambient daylight is calibrated against the gray side of the Kodak Neutral Grey Card (KGC) with 20% reflectance. (Winkler 1979)

exposure value (EV), given for the Kodak Grey Card and the object. The following equation can be used:

reflectance of sample/reflectance of Kodak Gray Card (18%)
$$= 2^{EV} \text{ change.}$$

Specifications may be readily established for allowable maxima of change.

Polishing a stone surface decreases the light value by darkening the color. A wet stone surface also decreases the value. Dark-colored stone varieties reduce the value more when polished than do light-colored stones. Sandblasted inscriptions on tombstones contrast the polished surface well, especially in red granites or dark gray to black gabbros (black granites). The curve of Fig. 4.7 plots the lighter versus the darker readings. Values were used to measure and plot reflections of sanded versus polished stone finishes.

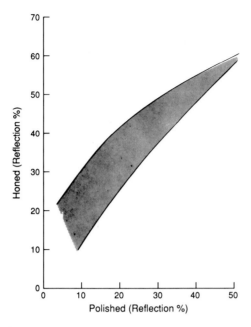

**Fig. 4.7.** Light reflection difference between honed and polished stone surfaces. The darker polished stone surface is averaged on the *lower curve*; the lighter honed and sanded surface is presented on the *upper curve*. The difference narrows toward stones of greater reflectance, the light-colored varieties. Data were measured with a Soligor spot meter on about 100 tombstones of mostly granitic rocks, with polished and sanded areas on a single stone block

## 4.2 Colors of Minerals and Rocks

The variety of mineral colors determines the color of the rocks. The $Fe^{2+}$ (ferrous) and $Fe^{3+}$ (ferric) forms of iron are the most important coloring agents in all rocks; organic carbon can have a range of gray to black in sediments. The color identification chart of Table 4.3 attempts an explanation of stone colors.

## 4.3 Color of Igneous Rocks

The origin and mineral composition of igneous rocks is described in Section 1.1. The colors of the rock-forming minerals are summarized in Fig. 1.2 and discussed in the following.

*Orthoclase* in granite and syenite: The common orthoclase feldspar group may display colors from deep flesh, pink to white. The abundance of

**Table 4.3.** Colors of minerals and rocks with percent age ranges[a]

| Color | Minerals | Igneous (%) | Sedimentary (%) | Metamorphic (%) |
|---|---|---|---|---|
| Red, pink | Orthoclase | Granite (50–80) | Sdst (0–15) | Gneiss (10–80) |
| | Hematite | [occurs in most rocks (1–5)] | | |
| | Calcite | | (0–100) | Marble |
| White, light | Orthoclase | Granite (50–80) | Arkose (0–15) | Gneiss (5–35) |
| gray, | Plag | Granite (5–50) | Arkose (0–15) | Gneiss (5–35) |
| glassy | Quartz | Granite (5–35) | Sdst (5–100) | Gneiss (5–70) |
| | | | Quartzite (100) | Quartzite (100) |
| | Muscovite | Granite (0–5) | Sdst (0–5) | Marble, schist, quartzite (0–10) |
| | Calcite | | Ls | Marble (90–100) |
| Gray (dark) | Plag | Diorite (50–80) | Sdst (0–10) | Gneiss (5–50) |
| Silver | Graphite | | Sdst (0–5) | Marble, slate (0–5) |
| gray | Coaly matter | | Sdst (0–5) | |
| | | | Shale, ls(0–5) | |
| Black | Hbl | Granite (5–10) | | Gneiss (5–50) |
| | | Biorite (10–25) | Sdst (0–5) | |
| | | Gabbro (20–60) | | Gneiss (20–60) |
| | Biotite | Granite (0–5) | | Gneiss (5–90) |
| | | Diorite, gabbro | | |
| | Graphite | | Sdst (0–2) | Marble, slate |
| | | | | Gneiss (0–10) |
| | Organic substances | | Sdst, ls | |
| | | | Shale (0–5) | |
| Brown | Hbl | | | Gneiss, schist |
| | Biotite | | | (5–80) |
| Green | Hbl | | | Gneiss, schist (0–100) |
| | Serpentine | | | Serpentines (Verde Antiques) (80–100) |
| | Ferrous-ferric | | Shales, ls, sdst (1–5) | Slate (1–5) |
| Yellow, buff limonite(rust) | | | Sdst, shale | |
| Tan ochre | | | Limestone (1–3) | |
| Gold | Sericite mica | | Ls (1–5) | Marble (1–5) |

[a] Sdst, Sandstone; hbl, hornblende; ls, limestone; plag, plagioclase.

orthoclase in true granites and syenites (50–85%) determines the overall color of these rocks. Local crystals of white orthoclase can give a spotty appearance which may stem from a previous intrusive generation. The origin of the red and pink hues is attributed to finely disseminated flakes of ferric oxide (hematite).

*Plagioclase* feldspars in diorites and gabbros: The plagioclase group presents a mixture series from the white sodium feldspar, through gray to the often black calcium feldspar. Plagioclase, in contrast to orthoclase, attracts the gray to black ferrous form of iron. The increasing iron content in dark igneous rocks provides enough iron to color the plagioclase black. Gabbros or "black granites" owe their often uniform and deep black color to black hornblende and black plagioclase. The dark labradorite of some black granites may show irridescence caused by light refraction along numerous narrow twinning planes. Labradorite as individual crystals is white to light gray when not associated with dark igneous rocks.

*Hornblende and augite* are common members of the large iron–magnesium silicate group, with colors ranging from green to black, rarely light green. The minerals tend to oxidize by first leaching out of the iron, when water enters, and then converting to a brownish or ochre-brown lustreless clay substance.

*Biotite* is dark brown to black from the presence of ferrous-ferric iron. Igneous rocks often contain biotite, which replaces the equally black horn-blende. The iron is only loosely attached to the crystal lattice. It leaches out readily, oxidizing to a brown ferric iron that discolors the stone surface to an ochre brown, or leaves rusty blotches. The bleached biotite tends to appear gold or silvery white.

*Muscovite* is colorless to silver gray and is of stable color. The mineral is uncommon, except in pegmatites and some sandstones.

*Quartz* in granites appears usually glassy gray, pearly, and occasionally as bluish opalescent grains. A color contrast may occur with other minerals because quartz tends to be translucent. The white or gray color of quartz is caused by entrapped air bubbles or imperfections of the crystal lattice. A blue color is caused by traces of titanium.

*Pyrite*, or fool's gold, is a common accessory mineral in all rocks. The yellow-brass to gold color dulls soon after exposure to the atmosphere and becomes ochre-brown, often combined with halos by sulfuric acid leaching (Sect. 9.3).

In summary, the pigments of the rock-forming minerals are very stable in igneous rocks. All minerals polish well and resist the deleterious city atmosphere, except for biotite and pyrite.

## 4.4 Colors of Sedimentary Rocks

The unlimited variety of colors and shades of sedimentary rocks challenges the architect and decorator. The soft warm colors are mostly caused by the concentration and degree of iron oxidation to ferric oxides, a most important and powerful pigment. Organic substances can produce a variety of grays to black. Some stable color pigments are inherited from igneous rocks, such as quartz and feldspars, when these minerals survive the processes of weathering and sedimentation. Other rock-forming minerals, such as clay minerals in mudstones and shales, are formed by weathering. The color of pure clays ranges from dull white to gray.

*Iron* is the most common and the strongest pigment in sedimentary rocks. It can cause a range of colors from deep red to orange, yellow, brown, or tan, to green, blue, and black. The variety of color is a function of the amount and degree of oxidation. Most of the iron is precipitated, introduced in the past as soluble iron or freed by bacterial action. Some of the iron may have been introduced during lithification and some during mountain-building processes. The iron occurs in the oxidized ferric ($Fe^{3+}$) or ferrous ($Fe^{2+}$) forms.

*Ferric iron* may be found as the red hematite ($Fe_2O_3$), the reddish-brown to brown goethite (FeOOH), and the mostly amorphous $Fe_2O_3 \cdot nH_2O$, common rust. The color of most sedimentary rocks and soils is derived from various quantities of mostly amorphous iron hydroxide. Often only a fraction of a percent is necessary to add a warm tone.

Red sandstones: Colloidal absorption of iron hydroxide, as thin coatings on the surface of quartz grains, is found in many sandstones, as well as in modern sand dunes in desert environments. The red sandstones of the American Southwest, the brownstone of the northeastern USA, and the German Buntsandstein are the best-known examples of such red sandstones from ancient deserts. Iron can also be introduced as secondary cement. Many breccias, conglomerates, and sandstones are often cemented with reddish-brown intergrain cement. Van Houten (1969) distinguishes grains of black oxides, as pigments in matrix and cement.

Red limestone, though rare, is a desirable decorative stone. Most red limestones are believed to be secondary, but can also be formed in a shallow lagoon with sufficient influx of red lateritic soil from a former shoreline. If the pigment was introduced during catastrophic events, scars of former crushing should be visible (see Fig. 1.17). Processes of lithification have frequently attacked angular fragments by dissolution, rounding grain edges to poorly defined nodules. Color saturation of these nodules can be less intense than the color of the cement, or vice versa. The popular Italian Verona Red limestone-marble is a very common decorative veneer stone in

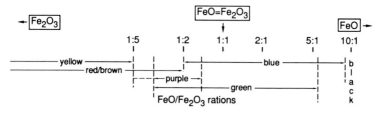

**Fig. 4.8.** Tanner's sediment color chart shows rock colors as they are related to the $FeO:Fe_2O_3$ ratio. (After Hughes 1958)

**Table 4.4.** Color of ferric:ferrous iron ratio

| Color (Munsell) | Total iron % | Fe$^3$ % | Fe$^2$ % | Fe$^3$:Fe$^2$ |
|---|---|---|---|---|
| Red part<br>10R3/4–10R4/6 | 0.22 | 0.09 | 0.13 | 1:1.5 |
| Gray part<br>5G6/1–5GY4/1 | 0.22 | 0.06 | 0.16 | 1:2.5 |

Color description from *Munsell Rock Color Chart*; Kieslinger (1964).

Europe and overseas. Strongly pigmented limestones with white calcite veins can also increase the beauty and contrast.

*Ferrous iron* may occur as finely distributed pigment that is bluish green to black in color, or as the yellowish-brass pyrite and marcasite. The power or streak of these minerals, however, is actually black. Figure 4.8 gives the colors of sedimentary rocks and metamorphic slates as a function of the ferric:ferrous iron ratio in the rock. Finely disseminated ferrous sulfide was formed in a reducing environment lacking oxygen. Gray or green reduction zones are common in limestones, shales, and slates. The iron proportions versus the color of Austrian Adnet limestone-marble, shown in Table 4.4, can be applied to most limestones.

Kieslinger (1964) demonstrates that the red and gray components of the Adnet limestone-marble contain the same amount of total iron, whereas the ferrous:ferric iron ratio changes. The sensitivity of the iron ratio is in evidence. The sediment color chart (Fig. 4.8) correlates the color as a function of the ratio of ferric to ferrous iron. Most sedimentary rocks have retained their original color during formation, reflecting their oxidizing or reducing environment. Such an environmental graph was sketched by Garrels and Christ (1965) for a range of pH and Eh, acidity versus oxidation-reduction (Fig. 4.9; see also Figs. 1.17 & 1.21). The diagram separates the stability boundary for $Fe^{3+}$ and $Fe^{2+}$, outlining most of the iron color stability in important environments. Verona Red, Rojo Alicante, and many other popular limestone-marbles are readily explained as follows: The marine limestone was probably deposited under slightly reducing conditions, later oxidized to bright red during regional tectonic activity, followed by subsidence where reducing waters penetrated along cracks, changing the red ferric iron to gray or greenish ferrous compounds (see Fig. 1.20). Soft shades of light green from traces of ferrous iron may spread in even distribution throughout the stone. Ferric iron solutions have traveled along cracks in the stone substance. Transmitted light displays a stunning array of colors, patterns, and shades for decorative objects and translucent windows.

*Ferrous carbonate*, siderite, can occur locally in limestones and crystalline marbles. Generally white or light cream, the iron carbonate dissociates,

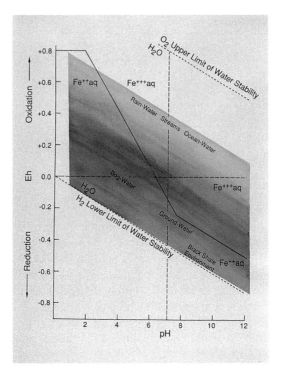

**Fig. 4.9.** Color of sedimentary rocks as a function of the pH and the Eh of the environment. (Modified from Garrels and Christ 1965)

though it is less soluble in acid rain than calcium carbonate, and immediately oxidizes to a yellow, then rust tarnish. The coating can be a continuous patina or local blotches. Kieslinger (1949) mentions the Greek Pentelian marble on the Acropolis of Athens as an example of yellowing of the stone surface by the possible presence of small quantities of siderite. Such staining is generally well developed along outer edges by the outward migration of capillary water, e.g., on vaporizing plates of home furnaces or on clay flowerpots (see Fig. 6.21).

*Glauconite*, a stable ferrous iron silicate of greenish color, is an occasional component of limestones and many green sandstones.

*Carbon*: Organic substance is frequently present in fine-grained sediments. These tend to oxidize when exposed to the atmosphere, the process occurring faster in coarse-grained sediments. The color of such sediments ranges from gray to deep black. Forsman and Hunt (1958) distinguish three types of organic matter in sedimentary rocks on the basis of their chemical composition:

a) Hydrocarbons – pure solvent-soluble organic matter composed of carbon and hydrogen only.

**Table 4.5.** Color and average organic content of sedimentary rocks (Hunt 1961)

| Rock | Color | Hydrocarbons (ppm) | Asphalt (ppm) | Kerogen (ppm) |
|------|-------|--------------------|---------------|----------------|
| Shales | Gray to black | 300 | 600 | 20 100 |
| Shales | Red and green | 15 | 40 | 1 000 |
| Carbonates | Gray to black | 340 | 400 | 2 160 |

b) Asphalt – solid and semisolid hydrocarbons largely soluble in carbon disulfide.

c) Kerogen – the bulk of the organic substance in shales. Insoluble in solvents, kerogen yields oil when the shales undergo destructive distillation. The carbon pigment of marine shales and limestones is closely related to the composition of coal. Hunt (1961) has calculated the average proportion of the three different types of organic materials enclosed in shales and limestones (Table 4.5).

The aromatic hydrocarbons, the least stable carbon pigment, determine the dark colors of sediments. The pigment stability of each carbon compound determines the expected color change during weathering. A relationship exists between the organic matter content and the degree of light reflection (Patnode 1941). Dark gray and black limestone-marbles have been popular for interior decoration since Baroque times. White fossil cross sections and white calcite fracture fillings make the stone still more desirable. Some dark gray carbonate rocks, such as some limestones and dolomites, may store and release $H_2S$ gas spontaneously when blasted or hit with a hammer. Some rocks may even change color from gray to pure white as the gas is released from the mineral grain boundaries.

## 4.5 Colors of Metamorphic Rocks

Gneiss, schist, slate, quartzite, and crystalline marble are the metamorphic rocks most frequently used.

The schistose or foliate structure gives these rocks a well-oriented pattern. Except for slates, the strong foliation of most metamorphic rocks rarely permits their use in the building trade.

Gneisses and most schists contain stable minerals inherited unchanged from their parent igneous and sedimentary rocks. The colors therefore resemble those of igneous rocks, except for slates and marbles. A few new pigment minerals are present, e.g., greenish chlorite, black graphite, and sericite, a very fine-grained muscovite.

Sedimentary limestones, sandstones, and conglomerates undergo considerable color change during metamorphism (Table 4.6). Pigment minerals

**Table 4.6.** Metamorphism of sedimentary pigments

| Sedimentary rock Pigment | Changes to Color | Pigment | Metamorphic rock Color |
|---|---|---|---|
| Limonite | Ochre, brown | Hematite | Red, black |
| Hematite | Red, reddish brown | Magnetite | Black |
| Soluble hydrocarbons | Gray to black | Burn out | – |
| Insoluble hydrocarbons | Gray to black | Graphite | Black to silver |
| Clay impurities | Shades of gray and green | Sericite, chlorite | Silver, white, gold, green |

recrystallize as other minerals do, and concentrate to form larger grains like all other minerals of the rock matrix. Frequently, the pigment mineral may occur as strongly concentrated bands, sometimes forming banded marbles of solid green or black. Carbonaceous limestones are metamorphosed to marbles in which carbon is concentrated as graphite in bands and along joints (Fig. 3.12). Warm tones of tan, buff, cream, ochre or black in the original sediments change to a cold white with green, silver, yellow or dark gray color. A finely distributed limonite in tan limestone-marble changes to a cold white marble with a few scattered black grains of hematite or magnetite. A dark gray or black limestone with carbonaceous pigment may change to a white marble with black flakes of graphite arranged in clusters or bands, or to an all-white marble if through dissolution the pigment mineral has migrated to another part of the rock. Hydrocarbons and pure asphalt are frequently distilled out during the process of metamorphism. Bain (1934) showed the transition of black fossiliferous bituminous limestone, the "Radio Black" of northern Vermont, into white crystalline marble of Proctor, Vermont, the Vermont marble of the Vermont marble-slate belt.

Light pink and red crystalline marbles are rare and the source of their pigment is not often clear, as is the case with Georgia pink marble, the Etowah Pink, or the Norwegian Rose. Red and pink quartzites, however, are quite common. While hematite colors quartzites, minute quantities of manganese appear to color marble pink. Tennessee marble is fragmental limestone and does not belong to metamorphic marble.

## 4.6 Special Color Effects in Stone

Special color effects in stone include the following:

1. Weathered joint surfaces and seam faces: Different color hues and chroma occur along joint surfaces where air and moisture have penetrated. Facings of freshly broken stone (split face) alternated with weathered faces can form a pleasingly colored surface. Discolored joint surfaces are neither

**Fig. 4.10.** Crossbanding of precipitated iron in concentric bands, Liesegang rings, are commonly found in Crab Orchard, Tennessee, sandstone. In the quarries the iron bands run parallel with the terrain surface. Retaining wall of planter, Morris Inn, Notre Dame University, Indiana

flaws nor a sign of stone weakness (Fig. 3.7). Such gentle color variations of the same stone, when set on retaining walls or bridges, break the visual monotony for the motorist and pedestrian. In contrast, a wild palette of colors with a frequent mismatch is generally considered poor design by architects.

2. Concentric band (Liesegang rings): Rhythmical concentric bands are well known in agates and were formed when they were in a gelatinous state. Liesegang's idea is described by Gore (1938), who demonstrates that ferric hydroxide may form rings by diffusion. Ochre-brown, yellowish brown to red rings of various thickness are found in some sandstones and weathered granites. Large field stones often show such weathering rings, which run approximately parallel with the original stone surface and open joints. Some sandstone quarries abound with brown bands of varied width intersecting the bedding planes (Fig. 4.10).

3. Stylolites: Dark irregular zigzag lines, stylolites, generally run parallel with the bedding planes in soluble sedimentary rocks. They are described and illustrated in Section 7.2. The greenish to black residual clay fillings of variable thickness may be responsible for special color effects. These are also lines of weakness and subject to weathering attack.

4. Dendrites: Branch-like thin coatings of brown to black iron-manganese films, dendrites, crystallized from solutions, are occasionally

deposited along joints and bedding planes; they resemble fine, moss-like or fern-like, feathery organic imprints. Solnhofen lithographic limestone of Germany is well known for its elaborate decorative dendrite formations on bedding planes.

5. Fossils: Cross sections of shells or coral branches in different orientation in an ancient shell bed or coral reef often make excellent decorative stones.

Decorative limestone-marbles (Fig. 2.22), ribbon slates with local bands of iron reduction from red to greenish bands (Fig. 1.24), should not be considered flaws but assets, unless the bands follow open cracks. The stone manufacturer should point out the potential variability of color irregularities to the customer; small sample slabs may become meaningless. The architect may have to select each panel if seeking certain combinations of color and texture. In general, metamorphic marbles are more consistent in color and fabric than sedimentary limestone-marbles.

Mosaics of differently colored and patterned stone pieces have been assembled since ancient times. The Notre Dame Library stone mosaic of Christ with his disciples (135 ft high and 65 ft wide), adorning the south face overlooking the football stadium, features 81 different kinds and finishes. A total of 5579 chips set against a frame and background of Mankato Gold dolomite offer the desired hue-chroma-lightness proportions (Winkler 1966). The pigment stability for each component had to be well known and the assurance given that later color changes would not disturb the well-planned balance laid out by the artist.

## 4.7 Color Stability of Minerals and Mineral Pigments

Yellowing of white marbles: With time many white marbles tend to discolor to cream, sometimes with a blotchy appearance. Small grains of biotite, hematite, or hornblende may lose iron, which immediately oxidizes. Only traces of ferric oxides are required to spread a buff "patina" over the marble surface.

Ferrous-to-ferric hydroxide: Many gray limestones, limestone-marbles, and dolomites tend to change color readily to yellow or cream on buildings and in quarries. In nature, the carbonate rocks exposed to the atmosphere show yellowish casts, while stone not exposed or only recently exposed remains gray or bluish-gray (Fig. 4.11). The color change can take place quickly within a few years. In contrast, the ochre-browns remain generally stable during exposure to the atmosphere. Joliet dolomite, Indiana limestone, and Mankato stone change readily from bluish to gold and cream colors. Mankato stone changes from "Mankato Blue" to "Mankato Gold" over decades. Pavement slabs of Mankato Blue at Notre Dame University

**Fig. 4.11.** Scaling of Joliet dolostone. The bluish gray of the original stone has changed to yellow-ochre at the stone surface by oxidation and hydration of the ferrous iron in the stone (see Fig. 4.9)

Memorial Library have discolored little where protected from rain, while the area exposed to rain has discolored to light cream.

Gray carbonaceous limestones and shales increase their value or lightness by oxidation and bleaching by the sun. Black limestone-marble panels on storefronts turn from deep black to pale "dirty" gray.

Black and green mica and hornblende are durable and can retain their color for 25 years or more.

Pyrite or "fool's gold" can change from the original sparkling gold to a flat and dull ochre-brown in a few years. Pyrite is formed and is stable in oxygen-poor environments (see Sect. 9.2).

# References

American National Standard (1985) Spectrophotometry and description of color in CIE 1931 system. Recommended practice for (ASTM E-308-66) Z-138.2, 1969, updated 1985

Bain GW (1934) Calcite marble. Econ Geol 29:121–139

Folk RL (1969) Toward greater precision in rock-color terminology. Geol Soc Am Bull 80(4):725–728

Escadafal R, Girard MC, Courault D (1989) Munsell soil color and soil reflectance in the visible spectral bands of Landsat MSS and TM data. Remote Sens Environ 27:37–46

Forsman JP, Hunt JM (1958/59) Insoluble organic matter (kerogen) in sedimentary rocks. Geochim Cosmochim Acta 15:170–182

Garrels RM, Christ CL (1965) Solutions, minerals and equilibria. Harper and Row, New York, 450 pp

Goddard EN et al. (1948) Rock color chart. Geol Soc Am Natl Res Counc, re-published by Geol Soc Am 1951

Gore V (1938) Liesegang rings in non-gelatinous media. Kolloidzeitschrift 32:203–207

Hughes RJ (1958) Kemper County geology. Miss State Geol Surv Bull 84:274

Hunt GM (1961/62) Distribution of hydrocarbons in sedimentary rocks. Geochim Cosmochim Acta 22:37–49

Judd DB, Wyszeski G (1963) Color in business, science and industry. Wiley, New York, 500 pp

Kelly KL, Judd DB (1976) Color, universal language and dictionary of names. US Dep Commer Natl Bur Stand Spec Publ 440:184

Kieslinger A (1949) Der Stein von St. Stephan. Herold, Vienna, 486 pp

Kieslinger A (1964) Die nutzbaren Steines Salzburg's. Bergland-Buch, Salzburg, 436 pp

Kodak (1963) Notes on practical densitometry. Kodak Salos Service Pamphlet E-59, Rochester, NY, 23 pp

Munsell Book of Color (1947) Glossy edition, Munsell Color Co. Munsell Color, 2441 North Calvert St, Baltimore, MD 21218

Munsell Soil Color Charts (1973) Munsell Color Co, Baltimore, MD

Nickerson Color Fan, maximum chroma, 40 hues (no date given), Munsell Color Co, Baltimore, MD

Nimeroff I (1968) Colorimetry. US Dep Comm Natl Bur Stand Monogr 104:47

Patnode HW (1941) Relation of organic matter to color of sedimentary rocks. Am Assoc Pet Geol Bull 25:1921–1933

Van Houten FB (1969) Iron oxides in red beds. Geol Soc Am Bull 79(4):399–416

Winkler EM (1966) Memorial Library, University of Notre Dame Stone Mosaic. Earth Sci 1966(2):56–58

Winkler EM (1979) The lightness (reflectance) of stone in the stone industry. Assoc Preserv Technol Bull XI(2):7–15

Wright WD (1969) The measurement of color. Hilger, London, 340 pp

# 5 Weathering Agents

## 5.1 Agents

The decay of stone and concrete in engineering structures and monuments
is closely related to the geologic process of rock weathering. Most of the
decay progresses near or at the ground surface, influenced by the following
weathering agents: atmosphere, rainwater, rising ground moisture, stream
water, lake water, and seawater. The atmosphere, rainwater, and rising
ground moisture, with or without dissolved salts, are most instrumental in
the decay of building and decorative stone. The influence of plants, bacteria,
and animals on stone decay is discussed in separate chapters.

## 5.2 General Considerations

The accelerated rate of decay of our cultural treasures in stone is becoming
a familiar sight. Figures 5.1 and 5.2 present the progress of decay; Fig. 5.1
shows time-lapse photos of a sandstone sculpture exposed to the industrialized Rhein-Ruhr area of northern Germany since 1702, photographed in
1908 and again in 1969. The weathering damage in the first 200 years was
relatively mild compared with that suffered from the beginning of this
century. Figure 5.2 shows more time-lapse cases from the area of Munich;
similar observations by Grimm and Schwarz (1985) are plotted into the
original curve by Winkler (1973), based on data of the statue shown in Fig.
5.1.

The process of weathering is an adjustment of the minerals and rocks to
conditions prevailing at the earth's surface, the present site of the building
or monument. The recrystallization of the original minerals to those of
greater regional stability is sketched in Fig. 5.3. The path of weathering
of rock-forming minerals in a particular environment of pressure and temperature of formation tends to move in a path, indicated by the arrows,
toward the weathering end product. The presence of oxygen leads to oxidation and the presence of moisture to hydration or to dissolution. The
readjustment may cause volumetric expansion of the crystal lattice, e.g.,
feldspars and the ferromagnesian silicates can hydrate to clays. Black micas,

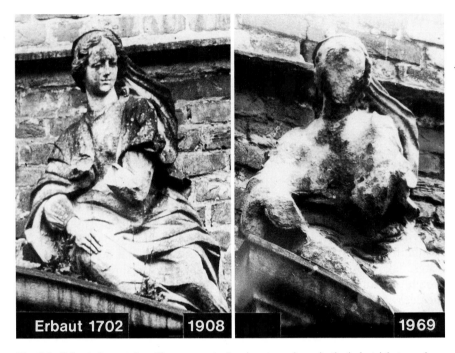

Erbaut 1702    1908                          1969

**Fig. 5.1.** Pair of photos taken 60 years apart, showing stone decay in the industrial atmosphere of the Rhein-Ruhr area. Sculpture in porous calcareous Baumberg sandstone, Herten Castle, Westphalia. Photos and information by Mr. Schmidt-Thomsen, Landesdenkmalamt Westfalen-Lippe, Münster, Germany

after rapid loss of iron and quartz, are very stable and remain in their original form for thousands of years or longer. Carbonate and sulfate minerals and rocks dissolve without residue. It is a slow geological process, yet it is rapid enough to inflict visible damage to a masonry wall in less than a decade. The environment of rock formation reflects the stability (Fig. 1.1).

The urban atmosphere of the twentieth century is creating special environmental problems to exposed stone surfaces. It can accelerate the process of decay to several times that in natural rural environs.

The chemical process of weathering is aided by mechanical break-up, which leads to the rapid enlargement of the mineral surfaces, providing accessibility to oxygen and moisture and resulting in accelerated destruction. The rate of destruction depends primarily on the climate. The appearance of weathered stone surfaces is not always a liability. Decorative boulders from old quarries in ancient glacial deposits often display differential coloring, bringing out the fabric not seen on freshly quarried material. It may provide character to both informal and formal foundation landscape planting and general landscape design. The following sections, which describe weathering agents, offer some detailed information on the corrosive urban environment conducive to stone decay.

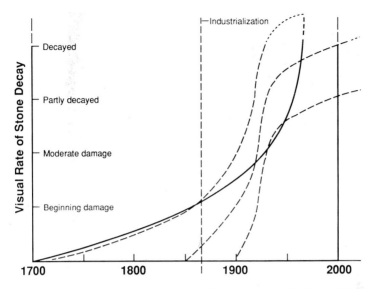

**Fig. 5.2.** Rates of stone decay, plotted from two different authors. Winkler (1973): time-lapse photos (Fig. 5.1), Herten; Grimm and Schwarz's (1985) curves are based on records of monuments in the area of Munich

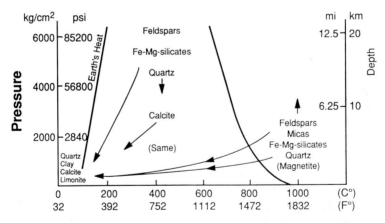

**Fig. 5.3.** Path of rock-forming minerals during weathering toward a new field of stability. (After Winkler 1973)

## 5.2.1 Atmosphere

The earth's atmosphere supports plant and animal life and has a basic composition of 78% nitrogen, 21% oxygen, and 1% $CO_2$ by volume; there are also traces of argon plus a number of impurities (pollutants), such as

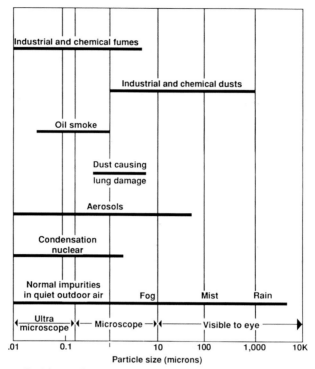

**Fig. 5.4.** Size distribution of important aerosols. (After EPA chart on Air Pollutants)

$SO_2$, $SO_3$, $NO_x$, $O_3$, $Cl_2$, and organic oxidants. These eventually settle out on the ground and on stone surfaces as aerosols (dry fallout). The role of this fallout is not always clearly evident. Aerosols with sizes ranging from molecules to raindrops (Fig. 5.4) usually settle with the rain and react with the stone substance in aqueous solution. The size range $10^{-2}$ to $10\,\mu m$ appears to be most important in the decay of stone. These are the larger ions, such as corrosive sulfates. While some sulfate, nitrate, and chloride is supplied in small quantities by natural causes, the majority is introduced into the atmosphere by automotive and industrial sources. Table 5.1 summarizes some of the components, their sources, and their effect on stone.

Continuous mixing of the air masses by winds and updrafts, and the partial neutralization of acids by suspended dust prevent the concentration of pollutant ions in the air to beyond human tolerance. Industrial pollutants have not yet penetrated the global atmosphere according to Fenn et al. (1963). Most of the pollutant ions return to the ground quite rapidly or are neutralized. The intensity of air pollution is reflected in the quantity of total suspended matter. The number and size of particles decrease with height but increase with rising relative humidity (RH).

**Table 5.1.** Important pollutants: sources and effects

|               | Natural sources                        | Industrial sources                          | Result of attack                          |
| ------------- | -------------------------------------- | ------------------------------------------- | ----------------------------------------- |
| *Particulates* | Volcanoes, forest fires               | Industrial exhausts                         | Soot cover                                |
| $CO_2$        | Volcanic, vegetation                   | Automotive, combustion                      | Acid rain                                 |
| $SO_2$        | Volcanic, desert dust                  | Combustion of fossil fuels                  | Acid rain and dry fallout                 |
| $SO_3$        |                                        | (Oxidation of $SO_2$)                       | Metals, plastics, acid rain               |
| $Cl$          | Volcanic, desert dust                  | Combustion, drycleaning                     | Acid rain                                 |
| $NO_X$        | Volcanic, vegetation, bacteria         | Automotive, combustion (stratosphere)       | Acid rain, photooxidation                 |
| $O_3$         | Trace in natural upper atmosphere      | Industrial, urban oxidation                 | Paints, plastics, metal corrosion, UV     |

**Table 5.2.** Concentration of ion-sized particles in the atmosphere

| Location                      | Average number per $m^3$ air |
| ----------------------------- | ---------------------------- |
| City                          | 147 000                      |
| Town                          | 34 000                       |
| Country: inland, seashore     | 9 500                        |
| Mountains   500–1000 m        | 6 000                        |
|     1000–2000 m | 2 130                  |
|     above 2000 m | 950                   |
| Islands                       | 9 200                        |
| Open oceans                   | 940                          |

After Junge (1958).

The chemical attack on stone is due largely to the solvent action of water and its dissolved impurities, like carbon dioxide, sulfate, and nitrate, inflicting acid corrosion. Table 5.2 gives the approximate distribution of ion-sized particles in the atmosphere for a variety of environments.

## 5.2.2 Solar Radiation

Solar radiation comprises short-wave UV, long-wave UV, the visible spectrum from violet to deep red, the invisible near infrared, and the far (thermal) infrared (Fig. 5.5). Much attention is paid to the "greenhouse effect": The visible spectral wavelengths of the sun penetrate the atmospheric gases, but not the clouds. Upon heating the land surface radiation reflects into space and is converted to much longer wave-lengths in the infrared (Fig. 5.6). Much of the radiation is absorbed by $CO_2$, and also by nitrate and ozone; most, however, is absorbed by the moisture in the atmosphere (Fig. 5.7). This reflected long-wave radiation cannot escape the earth's atmosphere, thereby warming the ground further. The gases function like the glass of a greenhouse, leaving heat inside.

### 5.2.3 Ultraviolet Radiation

Ultraviolet radiation (UV) is at the low part of the solar spectrum. It is invisible to the human eye, but destructive to organic consolidants and to the human skin and eyes (Fig. 5.5). UV radiation does not directly effect the stone substance. Organic consolidants and sealers, however, are usually attacked and lose their cohesiveness rapidly by disintegration. The yellowing of sealants and consolidants is the first indication of disintegration.

The radiation intensity is maximal near the equator, but diminishes rapidly toward the poles by the decreased angle of incidence. Figures 5.8 and 5.9 reflect the relationship between the radiation intensity and the maximum angle of incidence. The relative humidity (RH) can influence the heat household of the earth's surface, as shown in Figs. 5.10 and 5.11. Table 5.3 compares the solar radiation in cal/cm$^2$ of horizontal with vertical stone surfaces at various orientations for the city of Berlin near the 52nd parallel.

The travel path through the absorbing atmosphere decreases further the temperatures at higher latitudes (Fig. 5.8). The graph (Fig. 5.7) sketches the incoming radiation reaching the stone surface after it has been filtered through the atmosphere: about 50% is absorbed by water vapor and carbon

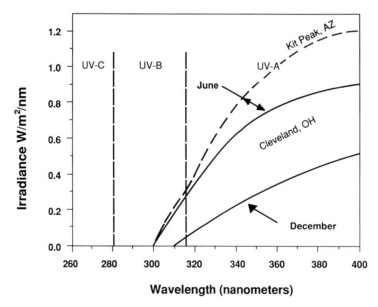

**Fig. 5.5.** Solar UV spectrum in mid-latitudes, Cleveland, OH (40°N). Note difference of UV radiation by season and elevation. (Adapted from Brennan and Fedor 1987)

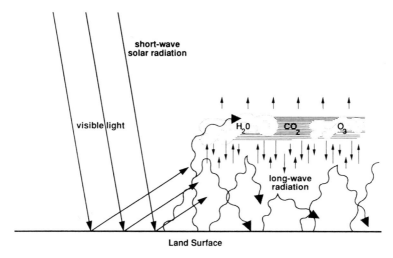

**Fig. 5.6.** Fate of visible solar radiation after warming and reflection (reradiation) into space, heavily filtered. (Adapted from Lutgens and Tarbuck 1979)

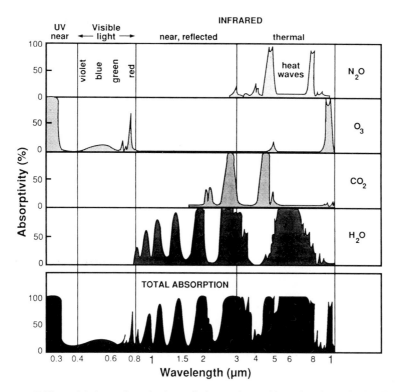

**Fig. 5.7.** Differential absorption of solar radiation at the earth's surface by $N_2O$, $O_3$, $CO_2$, and $H_2O$, by wavelength. Note the absorption of UV by ozone, and of heat waves by $N_2O$ and water. (Fleagle and Businger 1963)

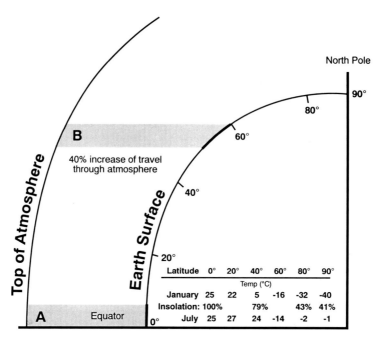

**Fig. 5.8.** Mean temperatures at latitudes and insolation received at various latitudes from the Equator (0°) to the North Pole (90°). Solar radiation tends to spread toward the poles and travel increases through the atmosphere. (Adapted from Powers 1966)

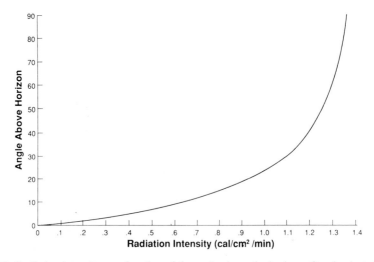

**Fig. 5.9.** Radiation intensity as a function of the angle above the horizon. Graph adapted from figures by Pogosyan (1962)

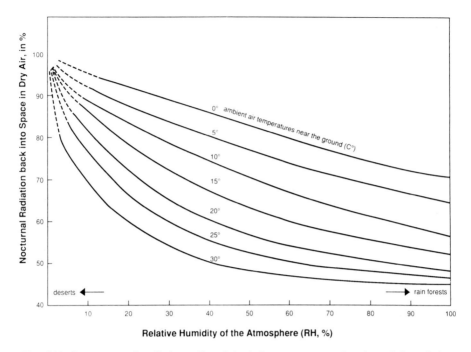

**Fig. 5.10.** Percentage of radiation reflected back into space as a function of the relative humidity (RH) for ambient air temperatures. Dry air permits maximum back radiation and maximum cooling, while moist air permits minimum back radiation and minimum cooling. Data from Geiger (1950)

dioxide, and some by ozone and nitrates. The absorption is distributed over a wide range, especially throughout the infrared, with

$$\begin{array}{ll} \text{water vapor:} & 5{-}7.5\,\mu\text{m} \\ \text{carbon dioxide:} & 12.9{-}17.1\,\mu\text{m} \\ \text{ozone:} & 9.4{-}9.8\,\mu\text{m}. \end{array}$$

Visible light with a maximum intensity at wavelength $0.5\,\mu$m (between green and blue) changes readily to longer wavelengths of near $10\,\mu$m as solar radiation hits the ground to a maximum intensity (Figs. 5.6 & 5.7). This radiation tends to be reflected readily into space in dry air after it has heated the ground, thereby generating great day-to-night temperature contrasts. Moist air absorbs radiation, causing minimum day-to-night cooling. The graphs of Fig. 5.11 characterize four major US cities with very different climatic characteristics, based on the mean day-to-night temperature differences with maximum relative humidity (RH) near the ground surface. The RH is inversely proportional to temperature contrasts and the nocturnal radiation back into space (Fig. 5.10). Relatively great RH can occur early in the morning, during which time moisture from the air can enter narrow

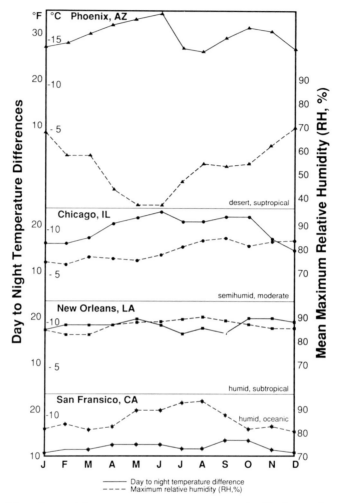

**Fig. 5.11.** Mean day-to-night temperature contrasts compared with the maximum relative humidity (RH). Temperature contrasts and the RH strongly influence the king and degree of weathering. Such plots may be drawn for any city in the USA and most places of the world. From data of NOAA (1980), National Weather Service, US Dept. of Commerce

stone capillaries. Such moisture enhances physical damage to stone by the large temperature range. The physical decay in dry climates contrasts the predominately chemical weathering in semihumid and humid climates. The greenhouse effect caused by $CO_2$ is probably minimal compared with the effect of moisture in the air.

**Table 5.3.** Solar radiation

| Total radiation | December (cal/cm$^2$) | Maximum radiation (cal/cm$^2$) |
|---|---|---|
| Horizontal surface | 15 | 319, June |
| South-facing wall | 65 | 158, April |
|  |  | 184, Sept. |
| East-facing wall | 10 | 148, June |
| West-facing wall | 12 | 139, June |
| North-facing wall | 0 | 23, June |

Solar radiation (cal/cm$^2$) for a mean day, computed under assumption for mean cloudiness for Berlin-Potsdam (Germany). Data from Schubert (1928).

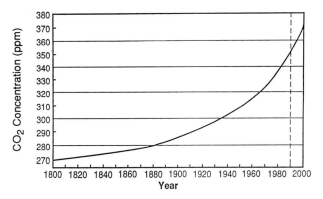

**Fig. 5.12.** Increase of the global $CO_2$ content of the rural atmosphere from 1800–2000. Seasonal fluctuations are smoothed out. (Modified from data of NOAA)

### 5.2.4 Carbon Dioxide

Carbon dioxide ($CO_2$) is a colorless, nonflammable, and nontoxic ingredient of the unpolluted atmosphere, 0.034% by volume. The atmospheric $CO_2$ increase may be traced to rapidly increasing industrial smoke and automotive exhaust. The relentless global deforestation means that forests cannot maintain a balance between production and absorption of $CO_2$, leading to an unavoidable, almost exponential, increase of $CO_2$ in the atmosphere. Figure 5.12 shows the fluctuations of $CO_2$ in the rural atmosphere since 1800. The increase of $CO_2$ from 0.0275% (?) in 1880 to near 0.035% in 1990 is believed to influence the gradual global warming through the greenhouse effect. $CO_2$ and other gases in the atmosphere have contributed only somewhat to this phenomenon, whereas water vapor remains the most powerful heat-absorbing factor (see Fig. 5.7). The absence of carbon dioxide absorption in caves may raise the level to 7% $CO_2$, a multiple of the surface atmosphere level; such concentration can readily attack cave paintings (Ek and Gewelt 1985). The two major sources of $CO_2$ are biospheric and industrial.

### 5.2.4.1 Biospheric Cycle

Organic disintegration, a part of the biospheric cycle, releases great volumes of $CO_2$. Photosynthesis by plants, on the other hand, consumes large quantities of $CO_2$ during daylight hours, when oxygen is produced. The process reverses during the night hours. The presence or absence of $CO_2$-emitting vegetation causes a seasonal sawtooth character of the $CO_2$ curve, smoothed for legibility (Fig. 5.12). The minimal annual increase of 0.2% is probably not sufficient to influence the temperature and the acidity of rainwater for the next hundred years.

### 5.2.4.2 Industrial and Automotive Sources

The rapidly increasing level of automotive exhaust is credited with more than 60% of all industrial exhausts. The caloric efficiency of combustion

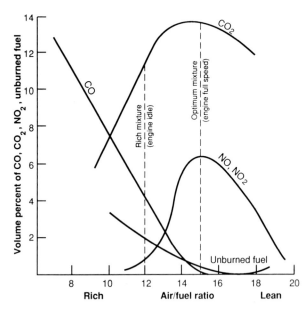

**Fig. 5.13.** Automotive exhaust: CO, $CO_2$, NO, $NO_2$, and the amount of unburned fuel at different fuel-to-air ratios. (Natl. Air Pollution Control Administration, Publ. AP-66, March, 1970)

depends mostly on the fuel-to-air ratio; an idling engine emits much CO and unburned hydrocarbons to produce a rich mixture (Fig. 5.13). Complete combustion at a 1:15 fuel-to-air ratio almost eliminates toxic CO and unburned hydrocarbons, with increased $CO_2$ and $O_2$. Sunlight in restricted basins, e.g., the Los Angeles basin, can convert unburned hydrocarbons to photochemical smog. The corrosive photochemical smog of the Los Angeles basin is a much-cited example. The nitrogen of the hydrocarbons oxidizes to $NO_2$ and then to corrosive nitric acid under the influence of strong oxidants, like ozone. Improved exhaust-pollution-control devices promise more efficient combustion in slow city traffic, eliminating toxic CO and unburned hydrocarbons; however, this raises the $CO_2$ level to almost 14% in the exhaust. The great volume of cars in developed countries will result in a continued rise in $CO_2$ unless we shift to a different means of transport, such as all-electric rapid transit with nuclear, solar, or hydroelectric generation of electricity.

## 5.2.5 Carbon Monoxide

Incomplete combustion emits millions of tons of carbon monoxide (CO) annually. CO is noncorrosive to stone or other materials but is very toxic to both man and animal: rapid depletion of the oxygen level in the blood leads to unconsciousness and death. CO does not oxidize readily to $CO_2$ in the atmosphere, but can act as a catalyst in the oxidation of $SO_2$ to $SO_3$. In urban areas the great concentration of CO during rush-hour traffic tends to reduce within a few hours. Soil bacteria in a healthy soil can metabolize CO and hydrocarbons rapidly (Abeles et al. 1971); good lawns in parks throughout urban areas can noticeably reduce the amount of CO and toxic unburned hydrocarbons.

## 5.2.6 Ozone

The triatomic form of oxygen ($O_3$), ozone, is a strong oxidant and a corrodent which tends to absorb UV radiation. Ozone is an important component of the atmosphere, though occurring in quantities less than 0.05 ppm. It is found in two separate layers, one of which forms a continuous envelope in the upper atmosphere 10 to 50 km above the land surface, where it absorbs most of the damaging solar UV radiation. Ozone in the lower atmosphere is concentrated in polluted urban areas, and fluctuates strongly with the time of day. It often exceeds 0.12 ppm early in the afternoon, depending on the wind speed, temperature, and solar radiation (Shreffler and Evans 1982). Figure 5.17 summarizes the ozone content of urban areas of the USA.

**Table 5.4.** Atmospheric sulfate contents. Data from EPA (1982; see Fig. 5.14)

| Location | Concentration ($\mu g/m^3$) | Environment |
|---|---|---|
| Rural England | 55 | Rural, marine |
| London (mean) | 285 | Urban |
| London (St. Pancreas) | | |
|    summer | 100 | Urban |
|    winter | 500 | Urban |
|    killing fog (1952) | 3840 | Urban |
| Pulp mill, New Hampshire | 5720–37 200 | Industrial |
| New York City, annual | 486 | Urban-industrial |
| New York City, 24-h aver. | 1085 (peak) | Urban-industrial |
| Kansas City, Missouri | 5.7 | Small urban |
| Chicago, annual | 286–542 | Urban-industrial |
| Denver, Colorado | 28.6 | Urban |
| Essen[a], Germany | 180 | Industrial |

[a] Anonymous (1969)

### 5.2.7 Sulfates, SO$_2$ and SO$_3$

Sulfate is the most powerful corrodent to most types of stone and materials. It usually appears as acid rain or dry fallout. The combustion of fossil fuels, coal, oil, and gas, releases sulfur as $SO_2$ into the atmosphere. Coal, the greatest offender, can contain as much as 8% total sulfur, whereas natural gas can have less than 2%. Sulfur in coal may occur as pyrite, the yellowish sulfide of iron, as the white gypsum, the sulfate of calcium, and as the invisible amorphous organic sulfur. Table 5.4 compares the sulfate content of the atmosphere at various places.

Strict antipollution laws have eliminated high-sulfur coal as a fuel in favor of lower total sulfur emission. The $SO_2$ gas can convert to the relatively weak sulfurous acid ($H_2SO_3$) in the presence of atmospheric water. The relative humidity (RH) of the atmosphere therefore strongly influences the amount of acid droplets from sulfur dioxide gas (Fig. 5.14). $SO_2$ can oxidize to $SO_3$ in smokestacks or in the atmosphere, aided by ozone and other oxidants and organic compounds found in the smog of urban industrial areas; this is illustrated for the USA in Fig. 5.17. The simplified chemical reactions are given for gaseous and aqueous reactions after Schwartz (1989):

$$\text{gaseous state: } SO_2 + H_2O = H_2SO_3 \text{ (sulfurous acid)}$$
$$\text{aqueous state: } SO_2\text{-}HSO_3 + H^+ = SO_3^{2-} + 2H^+$$
$$HSO_3^- + H_2O_2 = H_2SO_4 \text{ (sulfuric acid)}.$$

Caloric powerplants and roasting of sulfide ores in smelters emit huge quantities of $SO_2$ through tall smokestacks. Many square miles of barren ground near ore smelters are the eerie witness of the catastrophic results of excessive sulfate fallout and can be witnessed in the mining areas of

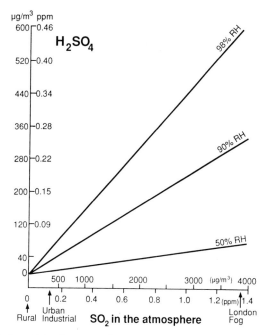

**Fig. 5.14.** Sulfuric acid in the atmosphere as an aerosol, as a function of the sulfate in the atmosphere and the relative humidity (RH). (Anonymous 1969)

Ducktown, Tennessee, and Sudbury, Ontario. Some of the sulfur in the atmosphere, however, finds its source in ocean spray and windblown dust picked up from desert floors. In rural areas some sulfate may also derive from the oxidition of $H_2S$ rising from the soil in small quantities in summer time. The scale for the height of the bars in Fig. 5.17 permits a close visual estimate of the degree of emission and a comparison with nitrogen dioxide and ozone. The heating of homes and offices in winter raises the seasonal sulfate output. Short, high peaks can occur at source-oriented sites. Luckat (1976) recorded emission rates of sulfate and chloride at different heights of urban buildings, e.g., the cathedrals of Cologne (urban industrial Rheinland) and Ulm (urban Bavaria). Chloride and sulfate show an increase at 45 m above the ground at Cologne (143% chloride, 131% sulfate), while in more rural Bavaria there is an increase with height only at 74 m (116% chloride, 127% sulfate). These figures conform with my own measurements taken at the base and on top (35 m) of the Field Museum of Natural History in Chicago, near the center of the city, surrounded by heavy traffic flow, with sulfate 150%, nitrate 193%, but no change in chloride. All data are based on 2-week summaries. Stronger weathering attack of coarse-grained Georgia marble on top of the Field Museum compared with attack at ground level is not surprising. A high RH leads to corrosive and often catastrophic acid fog (Figs. 5.14 & 5.15). Table 5.4 gives the sulfate contents of various places

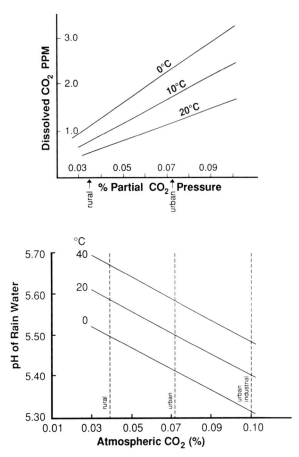

**Fig. 5.15.** Saturation of carbon dioxide in the atmosphere, in equilibrium with rainwater, and the pH of rainwater. (Winkler 1966; pH also calculated by this author)

and human environments. $SO_2$ can slowly oxidize to $SO_3$ in the atmosphere under conditions of intense sunlight and this occurs even faster in the presence of catalytic oxidants, such as ozone and high RH (Fig. 5.16).

   Sulfates can attack carbonate rocks in a dual way: by dissolution from the action of sulfuric or sulfurous acid, and by the change of carbonates to either calcium sulfate or calcium sulfite. Calcium sulfite has the solubility of calcite in pure water but is not influenced by either temperature or the $CO_2$ content of the solvent. Calcium sulfate, in contrast, is much more soluble than both calcite and calcium sulfite. The sulfates of magnesium are much more soluble than gypsum.

   The sulfate attack on silicate rocks, in contrast, is not easily measured. Leaching of alkali metals and/or iron may take place. Aerobic bacteria no doubt participate in the processes.

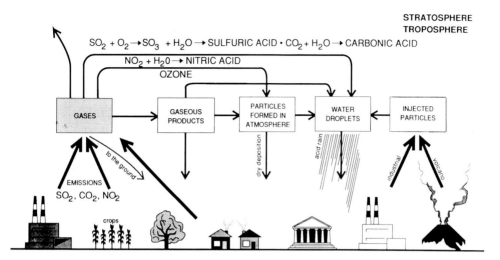

**Fig. 5.16.** Fate of pollutants, both natural and man-made. (Adapted from Graedel and Crutzen 1989)

### 5.2.8 Chloride

Chloride is an important constituent of the atmosphere and is emitted through industrial pollution. Some of the chloride may convert to hydrochloric acid, which readily dissolves carbonate rock. Much chloride is introduced as particulates as desert dust and ocean spray (Sect. 5.4).

### 5.2.9 Nitrate

The atmosphere is composed of 78% nitrogen. The oxides of nitrogen, however, may measure 2 to $4 \mu g/m^3$, with much higher concentrations registered in urban and urban-industrial areas (up to $400 \mu g/m^3$ in Los Angeles). Figure 5.17 gives a quantitative distribution of the nitrate emissions over the conterminous USA. Efficient automotive combustion (Fig. 5.13) is an important contributor of nitrogen oxides, which readily convert to corrosive nitric acid in the presence of oxidants. Though also a strong acid, nitric acid is less damaging to carbonate stone than sulfuric acid and sulfates due to the greater reactivity of sulfates with stone (Fig. 5.18).

## 5.3 Removal of Ingredients from the Atmosphere

Our atmosphere would be poisonous if the removal of pollutants from the air did not take place continuously. It keeps the level of toxic substances

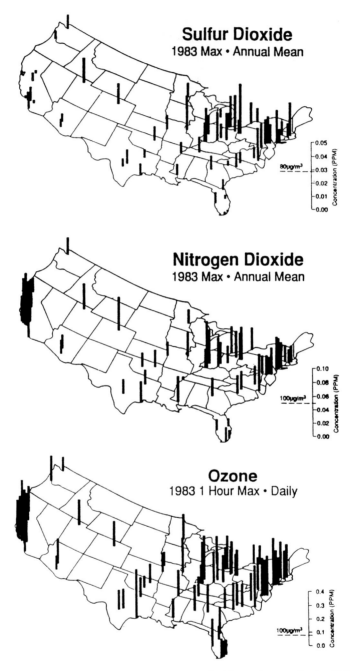

**Fig. 5.17.** Distribution of sulfur dioxide, nitrogen dioxide, and ozone in the United States, 1983. (EPA 1985)

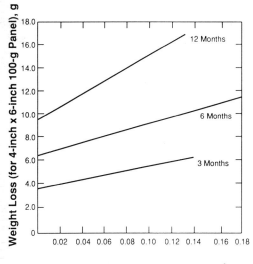

**Fig. 5.18.** Relationship between weight loss (corrosion) of mild steel plates and the $SO_2$ concentration of the atmosphere. (EPA 1982)

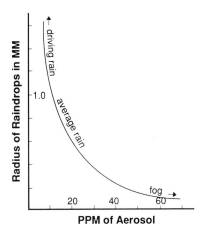

**Fig. 5.19.** Maximum adsorption of aerosols as a function of the raindrop diameter. (After Junge 1958)

below the upper limit of tolerance, despite greater levels within and near industrial and human centers. Junge (1958) considers several possibilities of removal:

1. Removal of aerosols:
   a) Dry fallout, as dust or soot.
   b) Washout of particles by precipitation (Fig. 5.19).

2. Removal of gases:
    a) Washout through adsorption by rain, the most important process (Fig. 5.19).
    b) Decomposition in the atmosphere by reaction with other gases, resulting in the formation of aerosols or smog.
    c) Absorption, by vegetation and stone decay.
    d) Possible escape of some light compounds into the outer atmosphere: carbon monoxide, carbon dioxide, nitrous oxide, and others.


## 5.4 Rainwater in Rural and Polluted Areas

The composition of rainwater reflects the composition of the atmosphere; the strong corrosive action is based on the ions collected during travel through the atmosphere. Whitehead and Feth (1964) distinguish rain, dry fallout, and bulk precipitation, which is the sum of both. The ions of rainwater show the marine effect and the composition of the atmosphere, whereas the dry fallout reflects mainly water-soluble contributions from local dust sources between rainfalls. Carroll (1962) discusses the acidity and ion content of rainwater and its influence on the chemical composition of surface waters. Rain may become very acid in areas of high sulfuric and nitric acids in the air, and is corrosive to stone, materials, and vegetation (Fig. 5.18). The quantity of dissolved $CO_2$ is essential as it influences the solubility of carbonate rocks and may accelerate the decomposition of silicate rocks. The pH in rainwater drops as the solution of $CO_2$ rises. Today, the term "acid rain" has become a household word. Rain is acid by nature with a theoretical mean pH of 5.6, based only on the present equilibrium of the $CO_2$ content in the atmosphere with rainwater, forming the weak carbonic acid (Fig. 5.15). Doubling or tripling of the atmospheric $CO_2$ from near 0.034 to 0.07% lowered the pH to 5.4, and at 0.1% $CO_2$ the pH was near 5.3.


### 5.4.1 Acid Rain

The pH of rain is not related to the greenhouse effect. Sulfates and nitrates in the atmosphere, however, can produce strong acids. Other natural contributions of sulfates and nitrates by volcanic eruptions, desert salts, and ocean spray can depress the pH far below 5.4 (Galloway et al. 1984). The amount of acid ingredients in the rain of remote areas versus industrial eastern North America is summarized in Table 5.5 (Fig. 5.18).

A distribution of the rain pH of the world is sketched in Fig. 5.20.

Aerosols can be removed from the air by the cleansing effect of the falling raindrops. The relationship of raindrop size and the amount of

**Table 5.5.** The pH and acid content of rain

|                          | pH        | Nitrates (ppm) | Sulfates (ppm) |
|--------------------------|-----------|----------------|----------------|
| Rural, remote[a]         | 5.0–4.7   | 0.16–0.31      | 0.1–0.45       |
| Industrial[b]            | 4.3–4.35  | 1.40–1.86      | 2.35–3.20      |

Figures adapted from Galloway et al. (1984).
[a] Remote areas, Bermuda, Australia, etc.
[b] Industrial eastern United States, Champaign, IL; State College, PA.

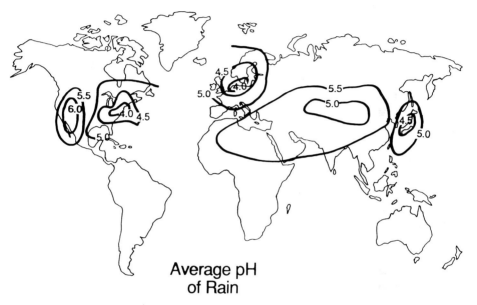

Average pH
of Rain

**Fig. 5.20.** World distribution of rain pH. *Contour lines* connect areas of the same rain pH. (Adapted from Park 1987)

absorbed aerosols is given in Fig. 5.19. The removal rate is determined by the height of fall from the cloud base to the land surface, the raindrop diameter, and the raindrop speed. The absorption equilibrium of rain is approached much faster by slowly falling small droplets with a larger specific surface area of absorption than by large drops of greater falling speed. Large raindrops can absorb a maximum of 10 ppm, whereas smaller fog droplets can absorb 60 ppm or more (Johnson et al. 1987). Urban winter fog can therefore be very corrosive and toxic, as was observed during the London fog of 1952, which killed over 4000 people (Figs. 5.14, 5.17 & 5.21).

### 5.4.2 Carbon Dioxide in Rainwater

Water can absorb gaseous ingredients, especially $CO_2$, twice as fast near $0\,°C$ than at $25\,°C$. Nordell's (1951) plot of $CO_2$ equilibria with rainwater in rural, urban, and urban-industrial atmospheres is shown in Fig. 5.15. The ordinate gives the $CO_2$ solubility in water.

### 5.4.3 Sulfate and Chloride in Rainwater

Sulfate is the strongest of the corrosive pollutants in rain because it can hydrolyze either to sulfurous acid from unoxidized $SO_2$ or to the more corrosive sulfuric acid from $SO_3$. The supply of chloride and sulfate is abundant near seashores and deserts as well as around highly industrial areas, such as the Chicago–Detroit–Cleveland region. The data in Table 5.5 present the pH range and the nitrate and sulfate contents of precipitation in both remote and industrial areas (Fig. 5.15).

## 5.5 Dust as Neutralizing Agent of Acid Rain

Dust is particulate matter in the atmosphere and ranges from 1 to $100\,\mu m$. Dust originates from both natural and industrial sources. Table 5.6 approximates the total world output of dust by origin.

Both anthropogenic and natural dust can affect the appearance and decay of stone.

**Table 5.6.** Sources of dust and total world output (After Robinson and Robbins 1971)

| Sources | Output, (millions of tons/year) | Percentage of total |
|---|---|---|
| Natural | | |
| Dust from soil, plowed fields, construction sites | 200 | 8 |
| Sulfates from $H_2S$ | 204 | 9 |
| Nitrates (electric storms) | 432 | 18 |
| Hydrocarbons, terpenes (forests) | 269 | 11 |
| Volcanic dust | 4 | Trace |
| Man-made | | |
| Particulates (aerosols), soot | 92 | 4 |
| Particalates (aerosols), sulfates | 147 | 6 |
| Particulates (aerosols), nitrates | 30 | 1 |
| Particulates (aerosols), hydrocarbons | 27 | 1 |
| Sea salt | 1000 | 42 |

### 5.5.1 Anthropogenic

Coke is generally derived from the burning of heavy fuel oils. Soot comes from most other fossil fuels. Soot tends to form a dense continuous coating on stone surfaces in urban and industrial areas. The formation and size of such particles is complex. The agglomeration of carbon atoms grows by complex chemical reactions, enabling these atoms to adsorb large quantities of gaseous hydrocarbons and other pollutants from the air. A coat of soot can form a tight film which can protect the stone surface while the interior of the stone decays. Carbonates may interact with absorbed sulfates to form accumulations of gypsum (see Fig. 7.5). The type of deposition determines the process of cleaning.

### 5.5.2 Natural Sources

While anthropogenic aerosols can be more destructive than natural sources, sea salts, windblown dust, and volcanic dusts are the most abundant sources. Soils and vegetation also contribute some sulfates and nitrates. Volcanic eruptions contribute sulfates and chlorides plus fragments of silicate glass.

Sea salt in the atmosphere is the greatest contributing factor. It is an abundant supplier of salts to masonry and materials up to a distance of 300 miles from coastal waters. Bursting bubbles of saltwater form water droplets of aerosol size; further reduction in size is caused by the evaporation of the water droplets.

#### 5.5.2.1 Windblown Dust

The total supply of windblown dust represents only 8% of all dust. The common minerals of natural dust are nonreactive quartz, clay minerals, calcite, feldspars, and ferromagnesian silicates. These minerals may help to absorb aerosols. Such dust reflects the area of origin. Limestone areas tend to produce calcite dust. Calcite reacts readily with sulfate in the atmosphere to form gypsum. The slow fall of dust through the atmosphere offers sufficient time for reactions to take place. A high relative humidity can speed the reaction time (Fig. 5.17). Gypsum is not of disadvantage to man. The lack of neutralizing calcite dust in New England exposes the area to acid rainwater with a pH near 4 (Likens and Borman 1974). Industrial dust removal devices can further diminish the neutralization of acid rain. The potential rain acidity raises the question whether strict pollution controls can reestablish preindustrial levels of rainwater acidity. Can the high acidity of rainwater caused by air pollution be neutralized by the greater production of neutralizing natural dust from construction sites, barren ground, and plowed fields (Winkler 1976)?

### 5.5.2.2 Volcanic Dust

The explosive eruption of volcanoes sends dust, primarily silicate glass, into the upper atmosphere, where it can travel for years before it settles back to the earth's surface. Little reaction occurs between volcanic dust and rainwater.

The degree of cleansing of polluted air by rain is difficult to assess. Meade (1969) estimates that the atmospheric contribution to dissolved stream load and lakes is 20% in North Carolina and 50% in New Hampshire after the reaction of rainwater with natural dust. The complexity of the interaction of the various natural and industrial factors is summarized in Fig. 5.16.

## 5.6 Stone Weathering
## Compared with Atmospheric Metal Corrosion

The corrosiveness of different atmospheric environments on metals is well known. The qualitative classification of corrosive atmospheres is dry inland or desert, semihumid or humid unpolluted rural, semirural, marine, and industrial marine. It appears that metal corrosion approaches the weathering of ferrous-ferric oxides and ferrous-ferric silicate minerals. Conductive minerals, like pyrite, marcasite, hematite, and magnetite, may also act as galvanic couples in the presence of an electrolyte. Such minerals may be subject to electrochemical corrosion by differential aeration. While dis-

**Table 5.7.** Corrosion rates of iron plates ($2 \times 4 \times \frac{1}{8}$ in.) in different atmospheres

| Location | Type of atmosphere | Latitude | Average weight loss (g/year) |
| --- | --- | --- | --- |
| Khartoum, Sudan | Dry, inland | 16° | 0.16 |
| Abisco, north Sweden | Humid, unpolluted | 69° | 0.46 |
| Singapore | Tropical marine | 2° | 1.36 |
| Daytona Beach, FL | Rural subtropical | 28° | 1.62 |
| Basra, Iraq | Dry inland | 31° | 1.39 |
| State College, PA | Rural, semihumid | 41° | 3.75 |
| Kure Beach, NC (800 ft from beach) | Marine, semihumid | 34° | 5.78 |
| Sandy Hook, NJ | Marine industrial | 40° | 7.34 |
| Pittsburgh, PA | Strongly industrial | 41° | 9.65 |
| Sheffield, England | Industrial polluted | 53° | 11.53 |
| Daytona Beach, FL (on beach) | Marine, rural | 28° | 20.43 |
| Kure Beach, NC (on beach) | Marine | 34° | 70.49 |

Table adapted from Larrabee and Mathay (1963). Weight loss and geographic latitude are given to correlate radiation and weight loss with the approximate environment and solar radiation (Fig. 5.17).

imilar metals are known to form electric cells, different stone types in direct contact are known to affect each other very differently by the entry of dissolved mineral matter into the adjacent stone. For example, limestone or marble leaching into granite leads to salt action. Figure 5.18 correlates the weight loss of iron plates with atmospheric $SO_2$ concentration. The values in Table 5.7 give an indication of the aggressiveness of different atmospheres. The intensity of solar radiation can be deduced from the geographic latitudes.

The future of air pollution and the corrosion of stone and metals will depend on the antipollution laws and action taken by man, such as increasing the efficiency of combustion engines and the development of alternate fuels. Such action may decrease the emission of pollutants and increase the potential life of building materials. Stone, the eternal building material, may once again deserve its name, "rock of ages".

## 5.7  Urban Climate

The urban climate is regarded as milder and moister than rural environs. Many weathering agents affect both climate and stone stability. The reflection of solar radiation against rows of houses in narrow streets causes snow to melt from urban areas first. Winds are generally less strong but may

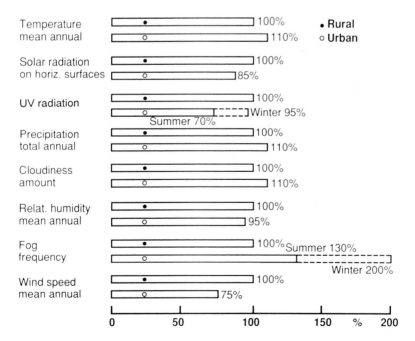

**Fig. 5.21.** Comparison of an urban with a rural climate. (Adapted from Lowry 1967)

become destructive in compression or suction on large window panels or stone-cladding panels; fog may be more frequent, especially in winter. Temperature, solar radiation, precipitation, cloud cover, relative humidity, fog, and wind speed between urban and rural conditions are contrasted in Fig. 5.21. This information is pertinent for the understanding of urban stone decay. The data presented in the chart are general. The distribution of residential housing, high-rise buildings, factories, park areas, and other urban properties play important roles in the final climatic evaluation of a site.

### 5.7.1 Temperature Variations

Temperatures average 10% higher, caused by numerous heat sources. The abundance of light-reflection surfaces can also help maintain higher

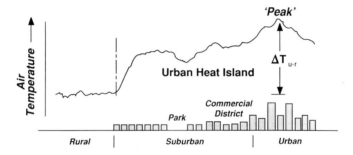

**Fig. 5.22.** Distribution of mean temperatures over an urban heat island relative to rural areas. T values are the difference between u and r (urban minus rural). (Oke 1978)

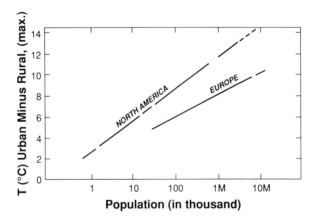

**Fig. 5.23.** Relationship of urban minus rural temperatures, versus the size of urban population, for the year 1970. Greater energy consumption per capita in North American cities influences the graph. (After Oke 1978)

temperatures despite the presence of smoke and haze, which can absorb up to 15% reflected light and infrared radiation. A typical urban heat dome is sketched in Fig. 5.22. The mean temperature increase from rural to urban is also a function of the size of the urban population. The increase of the air temperature with increasing population is greater in North America, because of the greater affluence, than it is in Europe (Fig. 5.23). Air temperature is one factor, but stone surface temperature is also an important contributor to stone decay (see Sect. 2.11; Figs. 2.13 & 2.14).

### 5.7.2 Cloud Cover

Precipitation is about 10% greater in urban areas, caused by the abundance of condensation nuclei from dust and smoke.

### 5.7.3 Relative Humidity (RH)

The RH is an important factor in the absorption of moisture into semiporous and porous stone. Urban humidity is about 5% lower than in rural areas as a result of lower average temperatures outside urban heat domes with the greater transpiration over agricultural lands and open water. This rural area

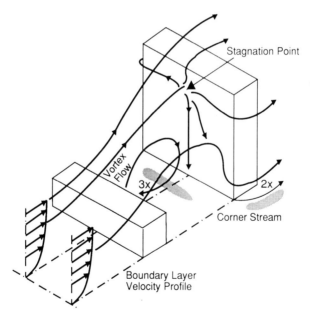

**Fig. 5.24.** Wind velocities and wind flow directions around urban obstacles. Such flow can accelerate the velocities by a multiple of the original wind velocity. (Adapted from Robinson and Baker 1975)

transpiration generally provides more moisture than in urban environments. Capillary moisture in stone from condensed water can inflict damage during heating–cooling cycles in urban heat domes, decreasing the strength of stone (Fig. 2.16) and reducing the effectiveness of stone consolidants. For more details, see Sections 6.1 and 6.2, and Figs. 6.2 and 6.8.

### 5.7.4 Fog

Fog is 30% greater in cities in the summer, but twice that amount in winter. Fog droplets can absorb pollutants to a high saturation capacity because of their large specific surfaces and very slow falling velocities. This increase in pollutants lowers the pH of fog, making it more acidic than rain. Limestone and marble columns often show deep erosion on the weather side, while retaining the original finish on the side protected from driving rain, despite the access of fog. Columns of Georgia marble in the Field Museum of Natural History in Chicago and columns of Verona Red limestone-marble at San Vitale in Ravenna have retained the original finish on the protected side for 70 and 1500 years, respectively (Fig. 7.8).

### 5.7.5 Wind Action

Wind action affects buildings directly. Although urban obstructions can reduce wind velocity by as much as 25%, winds can still exert both pressure and negative suction on building material. High winds can also generate driving rain, forcing water into masonry. Urban areas therefore compare unfavorably with their rural counterparts because moist surfaces cannot dry as quickly when 15% less drying sunlight is available. The reduction of wind speed in various urban, suburban, and rural environments is modeled in

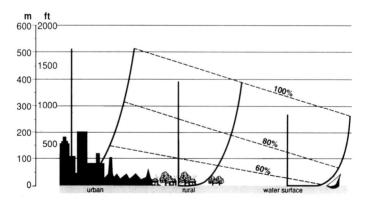

**Fig. 5.25.** Wind speed reduction determined by urban obstacles. Percentage of full force wind gradients. (After Davenport 1961)

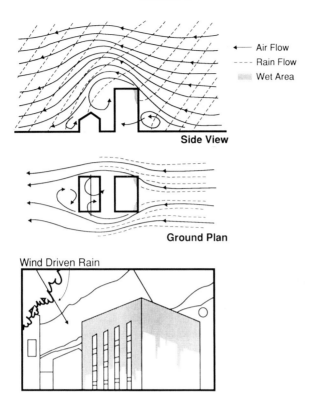

**Fig. 5.26.** Airflow over buildings with rain whipped against the building, shown in different views. Wetting of building after 40 min of directional exposure to rain. (After Robinson and Baker 1975)

Figs. 5.24 and 5.25. The airflow and formation of vortices often associated with increased wind velocities is sketched here following the principles of fluid mechanics (Davenport 1961). Thin, vertical stone-wall curtains should be carefully tested when used in zones of excessive compression or suction. Wind-driven rain forces moisture into the buildings. Robinson and Baker (1975) summarized the distribution of wetting on a building exposed to the vortices of wind-driven rain (Fig. 5.26).

Urban areas compare unfavorably with their rural counterparts: more moisture, less sunshine, and less drying winds.

# References

Abeles FB, Craker LE, Forrence LE, Leather GR (1971) Fate of air pollutants: Removal of ethylene, sulfur dioxide, and nitrogen dioxide by soil. Science 173:914–916

Anonymous (1969) Reinhaltung der Luft in Nordrhein-Westfalen. Rep Conf on Air Conservation in Düsseldorf, Oct 13–17, 1969, Ministry of Labor and Welfare, Nordrhein-Westfalen

Brennan P, Fedor C (1987) Sunlight, UV, and accelerated weathering. Plast Compound 12(1):44–49

Carroll D (1962) Rainwater as a chemical agent of geologic processes – a review. US Geol Surv Water Supply Pap 1535-G:1–18

Davenport AG (1961) The application of statistical concepts to the wind loading of structures. Proc Inst Civil Engin 19:6480

Ek C, Gewelt M (1985) Carbon dioxide in cave atmospheres, new results in Belgium and comparison with some other countries. Earth Surface Processes and Landforms 10:173–187

EPA (1982) Air quality criteria for particulate matter and sulfur oxides, vols I and III. Res Triangle Park, NC

EPA (1985) National air quality and emissions trends report. Office of Air Quality, Planning and Standards, EPA-450 4 84 029, 307 pp

Fenn RW, Gerber HF, Wasshausen D (1963) Measurements of the sulfur and ammonium component in the arctic aerosol of the Greenland ice cap. J Atmos Sci 20(5):466–468

Fleagle PG, Businger JA (1963) Introduction to atmospheric physics. Int Geophys Ser 5:346. Academic Press, London

Galloway JN, Likens GE, Hawley ME (1984) Acid precipitation: Natural anthropogenic components. Science 226:829–830

Geiger R (1950) The climate near the ground. Harvard Univ Press, Cambridge, MA, 482 pp (transl from German by Stewart MN)

Gorham E, Martin FB, Litzau JT (1984) Acid rain: Ionic correlation in the eastern United States, 1980–1981. Science 225:407–409

Graedel TE, Crutzen PJ (1989) The changing atmosphere. Sci Am 265/9:58–68

Grimm WD, Schwarz U (1985) Naturwerksteine und ihre Verwitterung an Münchener Bauten und Denkmälern. Überblick über eine Stadtkartierung. Bayer Landesamt Denkmalpflege 31:28–118, Munich

Johnson CA, Sigg L, Zobrist J (1987) Case studies on the chemical composition of fogwater: The influence of local gaseous emissions. Atmos Environ 21(11):2365–2374

Junge CE (1958) Atmospheric chemistry. In: Landsberg HE, van Miegham J (eds) Advances in geophysics. Academic Press, London 4(1):1–108

Larrabee LR, Mathay WL (1963) Iron and steel. In: La Quee FL, Copson HR (eds) Corrosion resistance of metals and alloys. Reinhold, Washington DC, pp 305–353

Likens GE, Borman FH (1974) Acid rain: a serious regional environmental problem. Science 184:1176–1179

Lowry WP (1967) The climate of cities. Sci Am 217:17–23

Luckat S (1976) Stone deterioration at the Cologne cathedral due to air pollution. In: Rossi-Manaresi R (ed) Conservation of stone I. Bologna, June 19–21, 1975, pp 37–43

Lutgens FK, Tarbuck ET (1979) The atmosphere, an introduction to meteorology. Prentice Hall, Englewood Cliffs, 413 pp

Meade RH (1969) Errors in using modern stream load data to estimate natural rates of denudation. Geol Soc Am Bull 80(7):1265–1274

Noll KE, Draftz R, Fang KYP (1987) The composition of atmospheric coarse particles at an urban and non-urban site. Atmos Environ 21(12):2717–2721

National Air Pollution Control Administration (1970) Publ AP-66

NOAA (1980) Climates of the States, 2 vols. Gale Res Co, Book Tower, Detroit

Nordell E (1951) Water treatment for industry and other uses. Reinhold, New York, 516 pp

Oke TR (1978) Boundary layer climates. Methuen, London, 372 pp

Park CC (1987) Acid rain, rhetoric and reality. Methuen, New York, 272 pp

Pogosyan KP (1962) The air envelope of the earth. GIMIZ Gidrometeorologicheskoe Izdatelstwo Leningrad, 1962 (transl by Izrael Program for Sci Transl 1965, 230 pp)

Powers WE (1966) Physical geography. Appleton-Century-Crofts, New York, 566 pp

Robinson E, Robbins RC (1971) Emissions, concentrations and fate of particulate atmospheric pollutants. In: Stoker HS, Seager SL (eds) Environmental chemistry: air and water pollution.

Final Rep March 1971, Stanford Res Inst Project SCC-8507, p 5, Scott & Foresman, 1976, 231 pp

Robinson G, Baker MC (1975) Wind-driven rain and buildings. Natl Res Counc Can, Division of Building Res, Tech Pap 445, July 1975, 19 pp

Schubert JD (1928) Sonnenstrahlung im mittleren Norddeutschland nach den Messungen in Potsdam. Meteorol Z 45:1–16

Schwartz SE (1989) Acid deposition unraveling a regional phenomenon. Science 243(4892): 753–763

Seinfeld JH (1989) Urban air pollution: State of the science. Science 243:745–752

Shreffler JH, Evans RB (1982) The surface ozone record from the regional air pollution study, 1975–1976. Atmos Environ 16(6):1311–1321

Waldman JM, Munger WJ, Jacob DJ, Flagan RC, Morgan JJ, Hoffmann MR (1982) Chemical composition of acid fog. Science 218:677–679

Whitehead HC, Feth JH (1964) Chemical composition of rain, dry fallout, and bulk precipitation at Menlo Park, California, 1957/58. J Geophys Res 69(16):3319–3331

Winkler EM (1966) Important agents of weathering for building and ornamental stone. Engin Geol 1(5):381–400

Winkler EM (1973) Stone: properties, durability in man's environment, 1st edn. Springer, Berlin Heidelberg New York, 250 pp

Winkler EM (1976) Natural dust and acid rain. Proc 1st Int Symp on Acid precipitation and the forest ecosystem, USDA Forest Serv Gen Tech Rep NE-23:209–217

# 6 Moisture and Salts in Stone

Moisture and salts are the most damaging factors in stone decay. The complexity of the capillary properties of a stone determines the hygric properties that are induced in buildings and monuments by moisture. The mechanism of moisture movement in stone in its different states is complex and not yet fully understood. Many field data are compiled from deserts and both humid and arid urban environments. Sources of moisture, moisture transfer in complex capillary systems of masonry walls, and the origin and action of salts are discussed for different climates.

## 6.1 Sources of Moisture

The important sources of moisture in masonry walls are moisture from rain, relative humidity (RH) in the atmosphere, rising ground moisture, and moisture from condensation near windows, doors, leaking pipes, roofs, and gutters.

### 6.1.1 Moisture from Rain

The relatively short duration of rain, even driving rain, and fog is too brief to permit moisture infiltration. Drying conditions extract most of the moisture following a rain. Surfaces of permeable sandstone, however, transmit moisture easily and allow it to penetrate deeper into the masonry. The important action is by washing and possible dissolution attack on masonry (Chap. 7).

### 6.1.2 Moisture from the Relative Humidity in the Atmosphere

1. Moisture can be either absorbed from wet surfaces if the atmosphere is undersaturated, or precipitated onto a colder masonry depending on the degree of atmospheric saturation. Adsorption of moisture into stone takes place in warm humid air when moisture comes into contact with cooler masonry walls or by cooling of the air temperature. Moist warm air readily

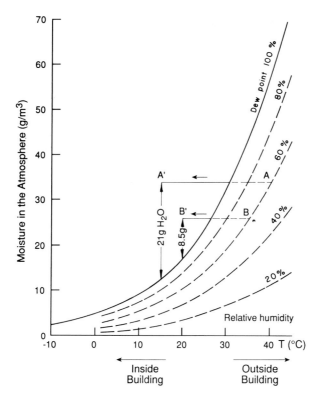

**Fig. 6.1.** Water content of air at 1000 mb pressure. Release of moisture is shown during cooling at constant moisture content: in *A-A'* cooling occurs from 40 to 15 °C along a wall, resulting in 21 g moisture freed. In *B-B'* only 8.5 g is freed when air is cooled from 35 to 20 °C. (Data adapted from US Weather Bureau)

enters the interior of a cold church or a building and condenses on the cold wall penetrating into the masonry, often traveling outward toward the surface, attracted by the sun and drying winds. The atmosphere can absorb considerable quantities of water depending on the ambient atmospheric temperature and pressure; the amount of freed moisture can be determined from the graph in Fig. 6.1. This theoretical model does not consider the effect of moving air. The data show the influence of high relative humidity as a major important factor in the supply of moisture into masonry, which is a function of the ambient RH of the atmosphere.

2. Moisture adsorbs readily to capillary walls in an oriented manner by electrostatic attraction in capillaries 0.1 $\mu$m or smaller. The positively charged $H^+$ ions of the water molecule are attracted to the negatively charged pore walls. The water molecules adhere tightly to the capillary walls with two, or possibly three, layers on each side of the capillary wall (Fig. 6.3). The charge decreases exponentially with increasing distance from the

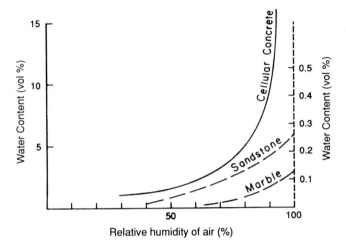

**Fig. 6.2.** Moisture absorption as a function of the relative humidity; data given for cellular concrete (Vos and Tammes 1969), sandstone, and marble from own laboratory tests. Also see Fig. 2.15

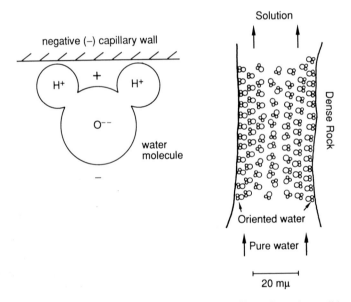

**Fig. 6.3.** Water molecule and "ordered" water in stone capillary. Osmotic conditions may be explained with this sketch

pore wall. The thickness of this film is 2 to 3 m$\mu$ at least 50% RH. The moisture sorption properties of fine-grained sandstone and slightly weathered marble are compared with cellular concrete in Fig. 6.2. Ordered water through capillary condensation can be instrumental for stone decay by moisture expansion, showing decreasing strength as a function of such condensation above freezing (Snethlage et al. 1988; see Fig. 2.15). Such ordered water is believed to be unfreezeable to near $-40\,°C$ (Dunn and Hudec 1966). The process of such disruption is not yet understood.

### 6.1.3 Ground Moisture

Ground moisture stems both from rain splashing against the base of a building and from groundwater rising by capillary suction. Both may contain more ions than surface waters, as salts from streets and sidewalks or from the slow circulation of undersaturated groundwater between the mineral particles of soils may be dissolved. As the water moves through the stone, the salts may crystallize and effloresce near or on the stone surface, or recycle back to the ground if they are soluble. Older buildings are rarely insulated against rising ground moisture. Section 5.2 provides more information on the ion supply of natural waters.

### 6.1.4 Leaking Pipes and Gutters

These are often hidden behind mortar and remain undetected, unless dark wet spots become visible on the wall. Condensation of moisture along plumbing and gutters may sometimes supply as much moisture by leakage.

### 6.1.5 Kitchen and Bathroom Showers

If not properly ventilated, kitchens and bathrooms provide a steady source of moisture which can travel toward the surface.

### 6.1.6 Construction Moisture

Construction moisture occurs if the ventilation of interior parts of a building was insufficient while the mortar and plaster were drying. Moisture problems in buildings are often very complex and should be identified and predicted in each case. Faulty insulation in modern buildings may be as bad as no insulation and moisture barriers in old buildings.

## 6.2 Capillary Travel and Capillary Suction of Moisture

The rise of moisture by capillarity exposes masonry walls and monuments to ground moisture. This can lead to slow and progressive salt impregnation. The maximum height of capillary rise is often astonishing, and is marked

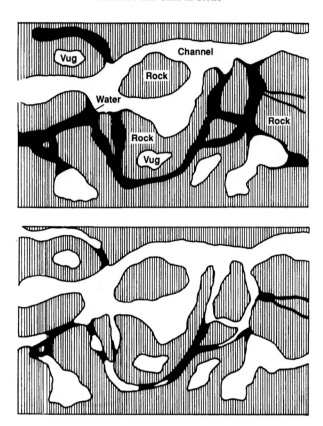

**Fig. 6.4.** Hypothetical pore system, applicable to dolostone. *Above* Continuous flow of water is possible through the pore system as only the narrower channels are filled. *Below* Water bubbles fill only part of the narrow passages between bubbles of vapor. Vapor locks prevent flow of water through the pore system

with a white rim of efflorescing salts with a wet outer rim. A hypothetical pore system in a fine-grained dolomite rock is sketched with different stages of water-filling in Fig. 6.4.

Fluid travel in masonry walls correlates well between field data and a simple equation of Kieslinger (1957) for capillary rise as a function of the pore diameter:

$$h = \frac{2s}{Rd},$$

where h is the height of capillary rise, s is a constant (about 0.074 g/cm), d is the density of water (1 g/cm$^3$), and R is the radius of the capillary.

The mean rise of water in capillaries of 1 mm diameter or smaller is sketched in Fig. 6.5. Pore diameter ranges of 20–6 $\mu$m (0.02–0.006 mm)

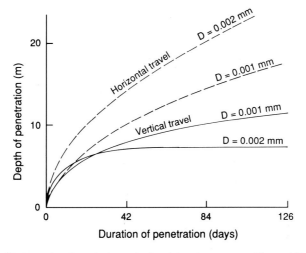

**Fig. 6.5.** Horizontal and vertical travel of moisture in masonry. Vos and Tammes (1969)

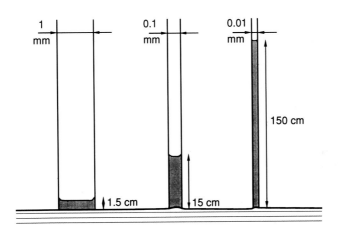

**Fig. 6.6.** Capillary rise of water in different diameter capillaries. (After Mamillan 1981)

should enable a maximum vertical capillary rise of 3–10 m; a diameter range of 6–2 μm (0.006–0.002 mm) permits 10–30 m rise, a great range (Fig. 6.6). Water travels twice as far in a horizontal as in a vertical direction. This probably explains the often strange and unexpected horizontal travel routes of moisture in old buildings. The composition and the wettability of the mineral matter is believed to have a strong influence on capillary rise. The vertical moisture distribution in a masonry wall resembles the moisture distribution in sand. The capillary rise of most of the water reaches only half the total height, to decrease to only 10% of the total water at the upper

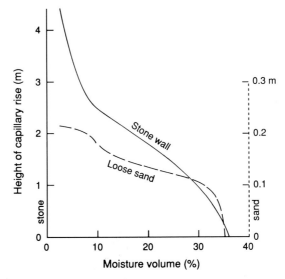

**Fig. 6.7.** Vertical moisture distribution in a stone wall (church of San Sebastiano, Venice) compared with moisture distribution in sand. (Adapted from Vos and Tammes 1969)

margin (Fig. 6.7). The maximum rise is often marked by a white efflorescent rim or a dark wet margin which remains wet by the hygroscopic effect of some salts. The moisture can be retained, especially in fine-grained masonry, stone, concrete, and mortar. Wind pressure, or a rapid change in barometric pressure, and the suction by evaporation through solar radiation and drying winds are additional factors to consider. Other complications may arise with the surface concealment of capillaries by lichens and mosses. Such an organic cover retains moisture and does not permit evaporation from the open ends of channel ways; consequently, the wall remains wet.

## 6.3 Capillary System

The capillary system is complex, and characterized by connected pore space of varying pore diameters. It can be measured with mercury porosimetry, water absorption, or image analysis from thin sections. Fitzner (1988) compares the pore size distribution of a German sandstone, fresh from the quarry, and a building, with the pore size distribution of weathered flakes from both locations. Pore size diameters of less than $0.1 \mu m$ from the weathered samples appear unchanged by the weathering process – such capillaries do not absorb moisture. Water absorption under atmospheric pressure amounts to about 50% of the total porosity (Fig. 6.8), with more absorption under vacuum. The weathering process shifts to larger pores by the gradual break-up of the grains, resembling crystallization tests. Salts in

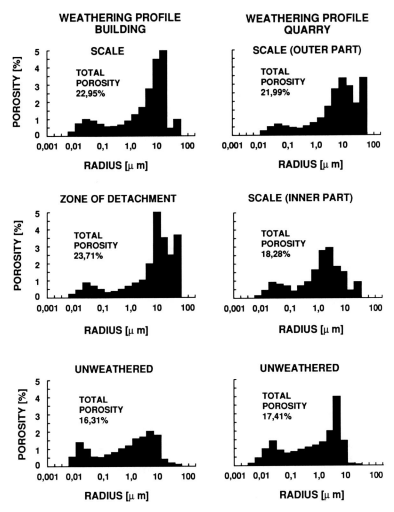

**Fig. 6.8.** Pore radius distribution of Schilfsandstein; weathering profiles of the sandstone both in the quarry and on a building. (Fitzner 1988)

pores increase considerably the adsorption of moisture from the air (Fig. 6.9).

## 6.4  Moisture Travel in Masonry

All moisture travels within the masonry. The travel through homogeneous pore spaces to the maximum height for a given mean pore space is relatively simple (see Figs. 6.5 & 6.9). Mortar joints and zones of denser stone can

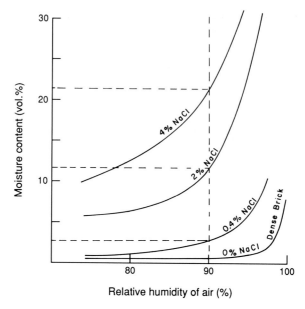

**Fig. 6.9.** Hygroscopic moisture in brick as a function of the relative humidity of the air and the salt content of the brick. (Vos and Tammes 1969)

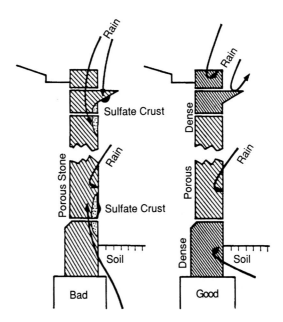

**Fig. 6.10.** Travel of moisture in a stone wall, with moisture both from rain and from the ground. Moisture flow of nonabsorbent stone complicates the flow pattern and the formation of crusts. (Mamillan 1968)

affect and change the travel routes. Mamillan (1968) sketches such water movement in stones of different density compared with a masonry wall of similar stone (Fig. 6.10). Porous mortar can even function as a pump, lifting water from one block to the next. Considerable travel heights can thus be reached.

## 6.5  Disruptive Effect of Pure Water

### 6.5.1  Stone Disruption by Wetting–Drying Cycles

Damage to stone by repeated wetting and drying is well known in stone decay studies. The term "hygric" is often used for stone decay by the action of water. Kessler and Hockman (1950) found that for granite rock expansion by moisture ranged from 0.0004 to 0.009%, with an average of 0.0039%; for quartz sandstone it was 0.01–0.044%. While we can model the disruptive action of heating–cooling cycles of trapped water, the disruptive action of "ordered water" is not yet understood.

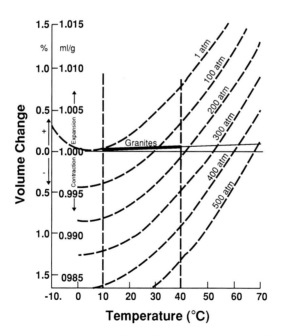

**Fig. 6.11.** Temperature – pressure – volume diagram of water above freezing. *Dashed lines* represent equal pressure water exerts when confined. Expansion is compared with granite. (After Winkler 1973)

### 6.5.2 Stone Disruption by Heating–Cooling Cycles

The temperature–pressure–volume change of water above freezing is considerable when compared with the thermal expansion of granite. A temperature range from 10 to 40 °C can exert 180 atm pressure against the capillary walls (Fig. 6.11). Ordinary water expands sufficiently to disrupt stone of low tensile strength when exposed to heating. For a temperature range of 30 °C granite expands 0.15%, while water expands 1.5% uncompressed. If the walls of small pores prevent water from expanding by 1.5%, water exerts about 380 atm (5600 psi) pressure against the pore walls of the granite. Light-colored igneous rock is not likely to reach surface temperatures of 60 °C, but dark igneous rock may reach even higher temperatures.

## 6.6 Case Hardening and Spalling

The sun and drying winds tend to provide a contrast of physical environments from the surface inward. Movement of moisture to the surface tends to redeposit binding mineral cement, dissolved from beneath the stone surface. Case hardening of stone is well known to the sculptor and architect, who "weather" stone on the beach or in the quarry before use. This method was practiced by Vitruvius nearly 2000 years ago, by Christopher Wren during the eighteenth century, and is still used today. Surfaces are hardened by quarry moisture as dissolved calcite or silica move from the inside of the stone to the surface, redepositing and strengthening the stone surface. Such case-

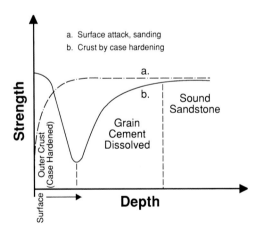

**Fig. 6.12.** Theoretical section through a block of sandstone in a building exposed to urban weathering. Sandstones tend to erode either (*a*) by surface sanding through loss of grain cement, or (*b*) by formation of a surface crust; grain cement dissolved behind the surface causes crumbling. (After Snethlage 1984)

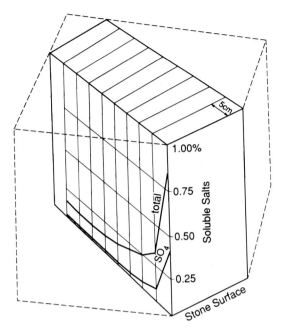

**Fig. 6.13.** Distribution of soluble salts across a green sandstone block, from the inside to the surface. The cornerstone was cut into slices 5 cm thick. The migration of salts is evident. Regensburg Cathedral. (Adapted from Kaiser 1929; Wendler et al. 1990)

hardened surfaces form a hardness profile (Fig. 6.12). Kaiser (1929) analyzed the movement of soluble salts to the surface of a dissected cornerstone from Regensburg cathedral (Fig. 6.13). Sculptors prefer to work on the soft, quarry-moist surface of Indiana-type limestone before the surface has hardened. In contrast to case-hardened surfaces, true crusts are secondary and often benignly seal the stone surface with a layer of soot mixed with airborne dust, often bound together with gypsum. Honeycomb formation is, in part, the result of interrupted case-hardened surfaces (see Sect. 6.13).

The material movement of dissolved mineral matter, calcite, silica, and iron, to the surface, often along a system of narrow spaced joints, leads to a strange boxwork (Fig. 6.14), similar to Kieslinger's (1959) frame weathering (Rahmenverwitterung) in paving stones. The frame is case hardened parallel with the sides of the stone at the cost of the cementing material. It has migrated from the center outward, leaving a less hard depression in the center; water tends to accumulate in this, leading to further decay. A spalled section from the weathered exterior of the Old Stone Church, Mission San Juan Capistrano, California, exemplifies the separation of shells parallel to the surface. The stone, sandy siltstone with a large clay fraction, was locally mined by the Spaniards about 200 years ago. Silica was leached from the fine clay fraction and has redeposited to form a dense surface layer (Fig. 6.15). This is usually found in sandstones with calcareous binding cement,

**Fig. 6.14.** Quartz, hematite, and limonite concentration along intersecting joints forming a boxwork. The sandstone crumbles between the joints to where the grain cement has migrated. The sandstone was exposed to a desert environment in the Nubian desert of Mt. Sinai, Egypt. Fifty years of exposure to semihumid South Bend, Indiana, as a monument has not changed the stone surface

but also occurs in porous limestones. A porous sandstone that has case hardened with time during 150 years of exposure to the climate is shown in Fig. 6.16; calcite-binding cement has migrated from the center part toward the outer edges, still preserving the details of the original tooling.

Boxwork is the result of similar migration of mineral substances outward to intersecting joints (Fig. 6.14). In the movement of calcite from mortar and cement, pointing and patching materials are important sources of Ca in calcite, Ca-bicarbonate, or Ca-sulfate moving into adjacent porous stone. Such movement can also occur from the concrete backing of veneer stone. Secondary crusts can form if the stone substance is dense. Porous limestones above a granite base course can provide enough calcite into semiporous

**Fig. 6.15.** Flaking (contour scaling) by salt action, mostly halite from neaby ocean spray. Soft sandy siltstone, surface hardened and scaling by migration of silica. Stone Mission Church, San Juan Capistrano, California. See also Fig. 6.13. (Wendler et al. 1990)

granite to cause flaking. Such combinations should be avoided, and mortar joints prevented from dissolution by rain and migrating moisture. In some small 4 by 4 in. stone blocks of noncalcareous sandstones on the National Bureau of Standards Stone Test Wall near Washington, DC, lime has migrated from the mortar into the stone surface (Fig. 6.17), causing extensive spalling over the entire block.

Flaking may also be induced by the treatment of a stone surface with a consolidant or a surface sealer when the boundary to the original untreated stone is abrupt. The depth of penetration is generally shallow. A gradual increase of surface hardness and density from the surface into the stone can prevent such flaking.

## 6.7 Origin and Transport of Salt in Masonry

### 6.7.1 Salts in Masonry

Natural stone may contain water-soluble salts entrapped in the pores. Most of these move invisibly in solution in the pores with the moisture, but others

**Fig. 6.16.** Weathering by case hardening. Tombstone of Aquia Creek sandstone; calcitic cement has moved to the upper surface, preserving the tool markings, while the stone is crumbling in the mid-section with a relief of about 25 mm. Stone from the Old Masonic Cemetery, Fredericksburg, Virginia. (Winkler 1977)

show at the surface as white or gray efflorescence. The crystallization in pores may disrupt the stone. Carbonates, chlorides, sulfates, nitrates, and oxides of iron are the most common constituents observed. The salts are found in a large variety of proportions, depending on their source. Salts of limited solubility show temporarily on stone surfaces in dry weather at low relative humidity. Salts with high solubility tend to remain in solution and rather attract more water; they are hygroscopic. Some strongly colored stones, especially those that are brown, red, or maroon, contrast efflorescent salts strongly.

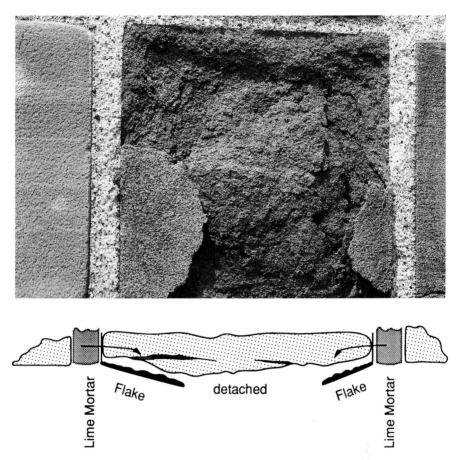

**Fig. 6.17.** Flaking of case-hardened sandstone. Lime mortar between the 4-in. blocks has traveled into the sandstone's surface layer following the usual profile (Fig. 6.12). A profile across the block is shown *below* the photo. Sandstone from Oregon, National Bureau of Standards Stone Test Wall, erected 1948 in Washington DC

### 6.7.2 Sources of Salts in Masonry

*6.7.2.1 Soluble Salts as Natural Constituents of the Stone*

New stone often shows white efflorescence of soluble sulfates or chlorides soon after removal from the quarry. Sulfates are common after natural weathering of pyrite.

*Calcium* dissolves from carbonate rocks and tends to form the sulfate gypsum in polluted air. Sulfates of Ca, Na, and Mg are most common among water-soluble salts in stone.

*Sodium* rarely exceeds 0.7% of the total stone substance; these salts lead to efflorescence and subflorescence in quarries and natural rock outcroppings. Sodium remains as a residue in the environment because vegetation does not absorb Na. Although rarely in excess, some geological formations are known to contain considerable amounts of soluble sulfates and chlorides.

*Iron* is a common constituent in the ferrous form, as the gray, bluish, or black ferrous form, from where it readily moves toward the surface, forming concentrations of brown oxidized ferric hydroxide. Iron and sulfur may also derive from grains or veins of pyrite. The process of decay may take only a few years.

### 6.7.2.2 Salts from the Ground

Groundwater is an ample source for salts, mostly as chlorides, sulfates, and nitrates of Na, Mg, Ca, P, and occasionally rare earth elements. Rising ground moisture is the vehicle of salt transport upward into the masonry. Organic pollutants can add phosphates and nitrates. The salts rising from the groundwater readily join salts leached from lime mortar, cement, concrete, plaster, and deicing salts on walkways.

### 6.7.2.3 Salts from Sea Spray

These salts occur more frequently and are carried much further than is generally believed. NaCl, the most common salt in the oceans, may be carried as fine spray up to 200 miles inland, settling on the ground or on top of buildings, from where it moves downward and upward. NaCl as well as some salts of Mg and Ca can form white rims. High winds blowing over the Great Salt Lake, Utah, have supplied ample quantities of NaCl to nearby Salt Lake City, infiltrating buildings and depositing salt on streets and cars. The decay of some granite blocks on the Mormon Temple show strong spalling as a result of salt action. The stones were mined as field stones in the now abandoned quarry at nearby Little Cotton Wood Canyon; these stones were previously exposed to natural weathering for many years. The old stone church of San Juan Capistrano in southern California shows extensive efflorescence and salt damage in the B-zone (Figs. 6.15, 6.19). The salt appears to be derived only from the ocean, a few miles distant.

### 6.7.2.4 Deicing Salts

In general use for only a few decades, deicing salt has done more damage to stone and concrete near ground level than any other destructive factor, as unnecessarily large quantities of NaCl, and more recently $CaCl_2$, are introduced onto walks and driveways.

### 6.7.2.5 Salts from Chemical Treatment

Some stone consolidants and sealants tend to release alkali metals during curing, especially Na and K waterglasses. Na and K tend to combine with sulfate, carbonate, and others to join the soluble salts traveling through the masonry. Among such ions the Na remains in the environment and recycles into the ground; it rises back into the masonry as it cannot be absorbed by plants.

### 6.7.2.6 Reactions of Salts in Old Walls

New compounds may form by the interaction of salts. Arnold and Zehnder (1990) describe reactions which may lead to new destructive compounds:

$$Na_2CO_3 + MgSO_4 = Na_2SO_4 + MgCO_3; \quad Na_2CO_3 + CaSO_4$$
$$= Na_2SO_4 + CaCO_3$$

$$K_2CO_3 + CaSO_4 = K_2CO_3 + CaCO_3; \quad K_2CO_3 + MgSO_4$$
$$= K_2SO_4 + MgCO_3$$

$$Na_2CO_3 + CaCl_2 = 2NaCl + CaCO_3; \quad Na_2CO_3 + MgCl_2$$
$$= 2NaCl + MgCO_3$$

$$K_2CO_3 + MgCl_2 = 2KCl + MgCO_3; \quad K_2CO_3 + CaCl_2$$
$$= 2KCl + CaCO_3.$$

The addition of deicing salts may revert some of the reactions given above.

## 6.8 Solubility and Salt in Masonry

Moisture and salts are drawn to the edges, from where water evaporates, leaving behind the salts as crusts or needles, like on evaporation plates of flowerpots (Fig. 6.18). A large variety of chemical compounds combines to form the salts entrapped in the stone pores, efflorescing at the stone surface. Arnold and Zehnder (1990) have recorded the vertical zoning of the salt distribution and action near the base of buildings. The salts rise from the ground level and are deposited according to their solubilities: the A-horizon, where only a few soluble salts crystallize, to the B-horizon, above the A-horizon, where most of the salts crystallize and inflict damage to the stone. Hygroscopic salts in the C-horizon leave wet masonry with little damage to

**Fig. 6.18.** Efflorescing salts, mostly NaCl, along the upper rim of a porous clay flower pot. Residual Na from the soil and fertilizers was rejected by the plant. Travel of soluble salts to outer surfaces, flakes, etc. is therefore by efflorescence

stone. In general, salts of low solubility tend to crystallize at an early stage near the ground surface, like most sulfates; salts of high solubility rise high and often remain in solution, forming wet zones (Fig. 6.19). The relative humidity of the surrounding atmosphere determines the temperature of crystallization at a given relative humidity. Table 6.1 includes the critical relative humidity at 20 °C for most common salts found in masonry.

## 6.9 Efflorescence and Subflorescence

Blotches and margins of white salts often deface stone and brick masonry, both near the ground and at levels high above the ground, occasionally beneath the roof line. Soluble halite and limited soluble gypsum crystallize

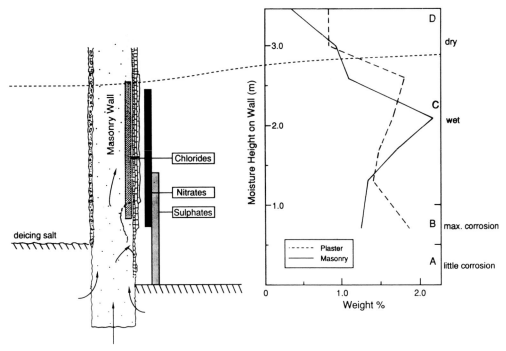

**Fig. 6.19.** Model of distribution of soluble salts in rising ground moisture on walls. (Adapted from Arnold and Zehnder 1990)

**Table 6.1.** Important soluble salts in masonry

| Compound | Solubility (H$_2$O) (at 20 °C, gm/100 ml) | Origin | Critical RH for stability (%) |
|---|---|---|---|
| MgCl$_2$·6H$_2$O (bischofite) | 167 | Mortar | 33 |
| CaCl$_2$·6H$_2$O ("ice-foe") | 279 | Deicing | 31 |
| K$_2$CO$_3$·2H$_2$O | 147 | Waterglass | 43 |
| Ca(NO$_3$)$_2$·4H$_2$O | 266 | Cement, groundwater | 54 |
| NaNO$_3$ (niter) | 91 | Groundwater | 75 |
| NaCl (halite) | 36 | Deicing, groundwater | 76 |
| Na$_2$SO$_4$ (thenardite) | 4.8 | Deicing, air pollution | 82 |
| MgSO$_4$·H$_2$O (kieserite) | 26 | Mortar, pollution | 65 |
| MgSO$_4$·7H$_2$O (epsomite) | 71 | Mortar, groundwater | 90 |
| Na$_2$CO$_3$·10H$_2$O (natron) | 22 | Deicing, groundwater | 98 |
| Na$_2$CO$_3$·7H$_2$O (heptahydrite) | 17 | Deicing | |
| Na$_2$CO$_3$ | 16 | Groundwater | 95 |
| Na$_2$SO$_4$·7H$_2$O | 20 | Deicing, groundwater | |
| Na$_2$SO$_4$·10H$_2$O (mirabilite) | 11 | Deicing, groundwater | 94 |
| CaSO$_4$ (anhydrite) | 2 | Not deserts | |
| CaSO$_4$·½H$_2$O (basanite) | Trace | Plaster-of-Paris | |
| CaSO$_4$·2H$_2$O (gypsum) | 2 | Mortar, pollution | |

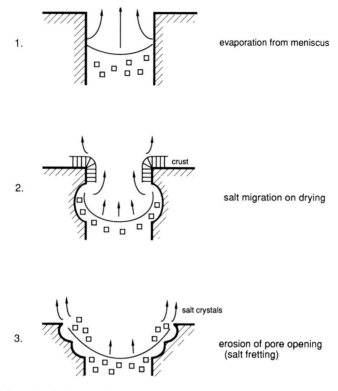

**Fig. 6.20.** Salt erosion by the creeping of salts by the mechanism of expansion-contraction, and hydration. (Adapted from Puehringer 1983)

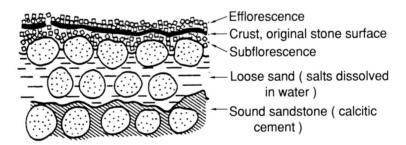

**Fig. 6.21.** Salt efflorescence and subflorescence, a schematized profile across an afflicted sandstone. (After Schmidt-Thomsen 1969)

at the surface as efflorescence, at the open ends of the capillaries or just beneath the surface as subflorescence; the open ends of the capillaries are closed by case hardening. Puehringer (1983) shows the moisture flow and salt deposition in a single capillary, as well as the corrosion inflicted within and at the end of the meniscus (Fig. 6.20). Changes of temperature and relative humidity may cause expansion and shrinking several times a day of the salt crystals connected with mechanical erosion. Upward creeping of salt crusts was also observed by Kwaad (1970).

*Subflorescence* is closely related with efflorescence. The salts on their way to the stone surface remain beneath the surface, crystallizing in the weathered rock substance behind an indurated crust. The skin eventually loses its support and starts peeling. Figure 6.21 sketches a cross section through a fine-grained sandstone which has developed a crust. It shows how salts can form subflorescence. Deserts and urban humid environments are favorable for salt action.

## 6.10  Damage by Salt Action

Salts are the most powerful weathering agent, especially when combined with frost action. Salt can act in several ways: by crystallization, hydration, differential thermal expansion, and osmosis. The complexity of damage by salt crystallization and hydration will be briefly discussed.

### 6.10.1  Crystallization Pressure

Expansion takes place when water freezes in a limited available space, e.g., a closed bottle may break or a car radiator rupture. The crystallization of salts from a supersaturated solution expands its volume and can crack stone and concrete. For many salts it is difficult to say whether the damage was done by pure crystallization pressure, hydration pressure, or thermal expansion of crystals. Nonhydrating salts, like halite, exclude damage by hydration. Evans (1969/1970) reviewed the literature on salt crystallization. Correns (1949) developed a workable equation for crystal growth pressure to quantify salt weathering in deserts, based on the Rieke principle: A crystal under linear pressure has a greater solubility than an unstressed crystal.

$$P = RT/V_S \times \ln C/C_S,$$

where P is the pressure by crystal growth (atm); R is the gas constant of the ideal gas law (0.082 l-atm per mol degree); T is the temperature (K); $V_S$ is the molecular volume of the solid salt; C is the actual concentration of the solute during crystallization; and $C_S$ is the concentration of the solute at saturation.

The state of supersaturation is essential for crystallization. Figure 6.22 plots the temperature T of the solution against the concentration ratio of the solution $C/C_S$ on an indefinite graph (Mullin 1961). Crystallization can only

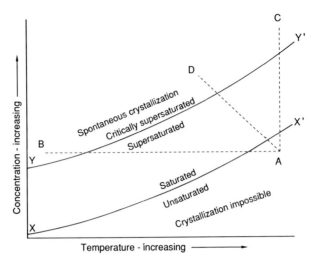

**Fig. 6.22.** Temperature – solute concentration plot, indefinite. Supersaturation can be obtained as follows: *A* to *B*, by cooling only; *A* to *C*, by an increase of concentration; or *A* to *D*, by a combination of *A* to *B* and *A* to *C*. (Modified from Mullin 1961)

occur between the states of saturation and supersaturation. No crystallization can take place in unsaturated solutions. Three theoretical situations are plotted on the graph. 1. *A to B*. The temperature drops at constant concentration from A toward B during a daily temperature cycle. Supersaturation is reached and crystallization can occur in this unstable state. With higher concentration of the solution, crystallization is more rapid. Lower concentrations and slower crystallization rates produce larger crystals. 2. *A to C*. The concentration increases at constant temperature by evaporation from the ends of capillaries. The process is reversible. 3. *A to D* . Change of both the temperature and the concentration leads to supersaturation where crystallization occurs.

Crystallization pressures were calculated by Winkler and Singer (1972) using Correns' original equation as the most practical thermodynamic approach with curves drawn for gypsum and halite. Sodium sulfate, the most efficient factor in rock disintegration, was also studied (Kwaad 1970). Figure 6.23 plots the log of supersaturation on the abscissa and the pressure generated by the crystal growth on the ordinate on a linear scale. The pressure–concentration relationship forms a straight line with the natural log of the supersaturation, $C/C_S$, for 0 and 25 °C. The curves are derived from theoretical calculations where conditions are considered ideal, though this is rarely found in nature. Table 6.2 summarizes calculations for salts which are common ingredients in rock pores, stone, and concrete. The pressures were calculated from the density, the molecular weight, and the molar volume for $C/C_S$ equals 2 and 10. Most calculated stresses are strong enough to disrupt most rocks (see Figs. 6.24 & 6.25).

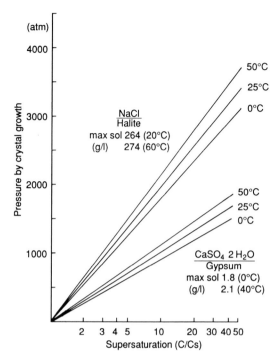

**Fig. 6.23.** Plot of crystallization pressures of gypsum and halite for various temperatures and degrees of supersaturation. (Winkler and Singer 1972)

## 6.11 Erosion by Deicing Salt

Salt action from deicing salts on drives and sidewalks has caused more damage to stone than any other source by adding salts to the environment. The saline solutions of NaCl, and more recently $CaCl_2$, disaggregate and flake porous and semiporous stone and concrete on walls in countries where deicing salting is practiced on streets and walks. The complexity and importance of the decay process deserves special attention. The salts can act in different ways:

1. By crystallization pressure, leading to crumbling and spalling. Figure 6.25 shows gray granite with clearly defined A, B, and C horizons. The B horizon is rough and white. Under the microscope the crumbling feldspars are not kaolinized, but are mechanically cracked and powdered. The presence of salt also increases the chemical dissolution attack. It has not attacked the feldspars. Halite (NaCl) is the most commonly used deicing salt to melt snow and ice (Table 6.3). Halite starts to crystallize at temperatures above 0.2 °C with a salt concentration of 36 g/100 ml water, but crystallizes as hydrohalite below 0.2 °C. Rising or falling tempera-

**Table 6.2.** Crystallization pressures of some salts

| Salt | Chemical formula | Density (g/cm³) | Molecular weight (g/mol) | Molar volume (cm³/mol) | Crystallization pressure (atm) | | | |
| | | | | | $C/C_S = 2$ | | $C/C_S = 10$ | |
| | | | | | 0°C | 50°C | 0°C | 50°C |
|---|---|---|---|---|---|---|---|---|
| Anhydrite | $CaSO_4$ | 2.96 | 136 | 46 | 335 | 398 | 1120 | 1325 |
| Bischofite | $MgCl_2 \cdot 6H_2O$ | 1.57 | 203 | 129 | 119 | 142 | 397 | 470 |
| Dodekahydrate | $MgSO_4 \cdot 12H_2O$ | 1.45 | 336 | 232 | 67 | 80 | 222 | 264 |
| Epsomite | $MgSO_4 \cdot 7H_2O$ | 1.68 | 246 | 147 | 105 | 125 | 350 | 415 |
| Gypsum | $CaSO_4 \cdot 2H_2O$ | 2.32 | 127 | 55 | 282 | 334 | 938 | 1110 |
| Halite | $NaCl$ | 2.17 | 59 | 28 | 554 | 654 | 1845 | 2190 |
| Heptahydrite | $Na_2CO_3 \cdot 7H_2O$ | 1.51 | 232 | 154 | 100 | 119 | 334 | 365 |
| Hexahydrite | $MgSO_4 \cdot 6H_2O$ | 1.75 | 228 | 130 | 118 | 141 | 395 | 469 |
| Kieserite | $MgSO_4 \cdot H_2O$ | 2.45 | 138 | 57 | 272 | 324 | 910 | 1079 |
| Mirabilite | $Na_2SO_4 \cdot 10H_2O$ | 1.46 | 322 | 220 | 72 | 83 | 234 | 277 |
| Natron | $Na_2CO_3 \cdot 10H_2O$ | 1.44 | 286 | 199 | 78 | 92 | 259 | 308 |
| Tachhydrite | $2MgCl_2 \cdot CaCl_2 \cdot 12H_2O$ | 1.66 | 514 | 310 | 50 | 59 | 166 | 198 |
| Thenardite | $Na_2SO_4$ | 2.68 | 142 | 53 | 292 | 345 | 970 | 1150 |
| Thermonatrite | $Na_2CO_3 \cdot H_2O$ | 2.25 | 124 | 55 | 280 | 333 | 935 | 1109 |

**Fig. 6.24.** Efflorescence of deicing salt, NaCl, on retaining wall of Indiana limestone. The line of efflorescence rises higher along pervious mortar joints. Note deep corrosion of the stone above the salt line, also strong scaling above the stairs. Indiana University at South Bend, Indiana

tures cause complete dissolution and crystallization from the solution without hydration. Such a process excludes hydration pressures (Braitsch 1962).

The crystallization of hydrohalite below freezing point can much accelerate the decay process of stone surfaces above street level, despite much lower crystallization pressures than for halite.

Iron can leach from ferromagnesian minerals, migrate, and redeposit near the border of zones A and B (Fig. 6.18), forming a light brown rim, sharp towards A and gradational into B. Similar ferric iron margins frequently appear in Indiana-type limestone.

2. Temperature shock through melting of ice. Klieger (1980) and also Roesli and Harnik (1980) discuss scaling by deicing salts, sodium chloride and calcium chloride, with the minimum temperature reached after only 1 min on a wet concrete floor. For deicer salt in solution 4% appears to be the optimum for scaling (Fig. 6.26). Semiporous stone is believed to behave similar to concrete used on walks, where scaling damage may be observed. Temperature shocks can also be effective on a vertical surface just above the ground level. The shock temperatures at shallow depths cause extensive internal compressive and tensile stresses, leading to spall-

**Fig. 6.25.** Deeply corroded granite by deicing salts. Flaking and surface reduction is highest in the center with a depression near 9 mm (see *arrow*). Travel of rising moisture is retarded along limited pervious mortar joints. Front entrance to the Hotel Intercontinental, Chicago, Illinois

ing on stone floors. Figure 6.27 shows deeply pitted Indiana-type limestone less than a foot above the heavily salted sidewalk. Two zones of pitting and crumbling are readily identified on the photo, with a lower and an upper zone only a few inches apart; the salt solution was apparently reflected by the impervious insulation to the bricks above.

Today, $CaCl_2$ is more frequently used for walks and stairways. It is more friendly to the environment, replacing the persistent Na with Ca that is readily absorbed by plants. The great solubility of the compound (279 g/100 ml) and a low critical RH of only 31% leaves the salt in solution after it has entered pervious masonry, forming lasting wet blotches (Fig. 6.28). The decay reaches deeper along the wet areas.

**Table 6.3.** Crystallization pressures of halite and hydrohalite

| Saturation | Pressure (atm) | | | | |
|---|---|---|---|---|---|
| | 20°C | 10°C | 0°C | −10°C | −20°C |
| *Halite* | | | | | |
| $C/C_S = 1.50$ | 353 | 341 | 328 | (318) | (306) |
| $C/C_S = 2.00$ | 595 | 574 | 552 | (533) | (515) |
| Solution | 36 | 36 | 36 | – | – |
| *Hydrohalite* | | | | | |
| $C/C_S = 1.50$ | – | – | 158 | 153 | 148 |
| $C/C_S = 2.00$ | – | – | 265 | 258 | 248 |
| Solution | | | 36 | 33 | 30 |

$C/C_S$ = degree of supersaturation common in nature; solution of salt per 100 g of water.
Data from unpublished notes.

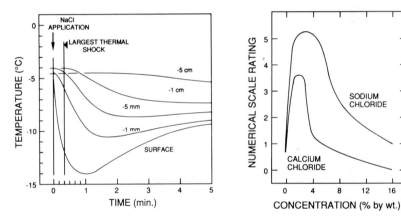

**Fig. 6.26.** *Left* Thermal shock as a function of time and temperature for halite. After Roesli and Harnik (1980). *Right* Scale rating as a function of the salt concentration for halite (NaCl) and CaCl₂. (After Klieger 1980)

## 6.11.1 Hydration Pressure

Some common salts readily hydrate and dehydrate in response to changes in the temperature and relative humidity of the atmosphere. The salt changes to a more stable hydrate, adjusting to the ambient environment. The absorption of water into the crystal lattice increases the volume of the salt, changing it to a new, more stable hydrate and exerting pressure against the pore walls. Mortensen (1933) first recognized the importance of salt hydration as a factor in desert weathering and attempted the calculation of hydration pressures. Monuments in cities of moderate humid and semihumid

**Fig. 6.27.** Damage of Indiana limestone by deicing salt. Note two horizons of corrosion, one just above the ground surface, and the other more intense one below the brick course with the insulating mortar as a reflector of the salt solution. *Numbers* on brick are the maximum depth of corrosion (mm). Grains of deicing salt are visible on pavement in front. Chicago, Illinois

climate may be readily saturated with salts introduced into masonry, as discussed before. Winkler and Wilhelm (1970) calculated the hydration pressures of some common salts at different temperatures and relative humidities. The following equation was developed:

$$P = \frac{nrT}{V_h - V_a} \times 2.3\log\frac{P_w}{P_{w'}},$$

where P is the hydration pressure (atm), n is the number of moles of water gained during hydration, r is the gas constant of the ideal gas law (82.07 mm-atm per mole-degree), T is the absolute temperature (K), $V_h$ is the volume of the hydrate ($cm^3$/g-mole of the hydrate salt, which equals the mole-weight of hydrate over the density), $V_a$ is the volume of the original salt before hydration ($cm^3$/g-mole; it equals the mole weight of the original salt over the density), $P_w$ is the vapor pressure of water (mm Hg) at a given temperature, and $P_{w'}$ is the vapor pressure of the hydrated salt (mm Hg) at a given temperature.

The pressure-temperature-humidity figures calculated in Table 6.4 are theoretical and present maxima under idealized conditions for closed pore systems. Indicated temperature stability ranges of hydrated minerals are

**Fig. 6.28.** Hygroscopic attraction of moisture by deicing salts, mostly CaCl$_2$. The wet areas remain visible throughout the year as the stone is strongly attacked. Balustrade is Indiana limestone; University of Notre Dame, Indiana

theoretical, as metastable conditions may exist for a long time at much lower temperatures than calculated. Calculations were performed with the sulfates of Ca, Na, and Mg, and the carbonate of Na. In general, low temperatures and high relative humidities produce the highest pressures, whereas high temperatures and low relative humidities produce low pressures.

The hydration of the hemihydrate plaster-of-Paris to gypsum takes only minutes, whereas anhydrate to the hemihydrate or to gypsum is a very slow geological process. Recrystallization does not always take place from one crystal lattice to another but passes through the liquid state. The anhydrite completely dissolves before it crystallizes to gypsum. The process is restricted to crystallization. No hydration pressure of anhydrite to gypsum can therefore take place. The process of hydration must be accomplished in a 12-h diurnal cycle to be effective by changes of temperature and relative humidity. The hydration pressure must also exceed the tensile strength of the rock. Snethlage et al. (1986) summarize the different types of salt weathering.

### 6.11.2 Differential Thermal Expansion

Natural mineralic building materials tend to expand at approximately 1 mm/m/100 °C. Salts enclosed in the pores may expand much more than the

**Table 6.4.** Hydration pressures of some important salts

| RH (%) | Temperature (°C) | | | |
|--------|------|------|------|------|
| | *CaSO$_4$·½H$_2$O to CaSO$_4$·2H$_2$O* (plaster-of-Paris to gypsum) | | | |
| | 0 | 20 | 40 | 60 |
| 100 | 2190 | 1755 | 1350 | 926 |
| 80 | 1820 | 1372 | 941 | 511 |
| 60 | 1375 | 884 | 422 | 0 |
| | *MgSO$_4$·H$_2$O to MgSO$_4$·6H$_2$O* (kieserite to hexahydrite) | | | |
| | 65.3 | | | |
| 100 | 418 | | | |
| 80 | 13 | | | |
| | *MgSO$_4$·6H$_2$O to MgSO$_4$·7H$_2$O* (hexahydrite to epsomite) | | | |
| | 10 | 20 | 30 | 40 |
| 100 | 146 | 117 | 92 | 86 |
| 80 | 115 | 87 | 59 | 39 |
| 60 | 76 | 45 | 17 | |
| 40 | 20 | | | |
| | *Na$_2$CO$_3$·H$_2$O to Na$_2$CO$_3$·7H$_2$O* (thermonatrite to heptahydrite) | | | |
| | 0 | 10 | 20 | 30 |
| 100 | 938 | 770 | 611 | 430 |
| 80 | 63 | 455 | 284 | 94 |
| 60 | 243 | 46 | | |
| | *Na$_2$CO$_3$·7H$_2$O to Na$_2$CO$_3$·10H$_2$O* (heptahydrite to natron) | | | |
| | 0 | 10 | 20 | 30 |
| 100 | 816 | 669 | 522 | 355 |
| 80 | 49 | 32 | 160 | |
| 60 | 60 | | | |

Data compiled from Winkler and Wilhelm (1970).

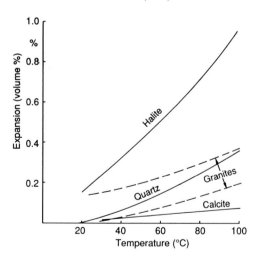

**Fig. 6.29.** Differential expansion of halite compared with quartz, calcite, and granite. (After Cooke and Smalley 1968)

stone, leading to thermal disruption if strong surface heating-cooling is available. Many salts obscure this process of thermal expansion with crystallization and hydration. Halite excludes the important hydration pressure. It was selected as an example in the graph shown in Fig. 6.29. The thermal expansion of halite is compared with the expansion of quartz, calcite, and granite. Upon temperature increase from freezing to 60 °C, halite expands 0.5%, whereas granite expands less than 0.2%. This difference appears small, yet disruptive expansion of salt can participate in physical stone decay. Rapid expansion by heating prevents possible plastic flow away from the pressure points. Snethlage et al. (1986) compare heat expansion with frost bursting (see Chap. 11).

## 6.12 Osmosis and Osmotic Pressure

Osmosis of salts through a solid occurs when a solution is separated from the pure solvent by a semipervious membrane, such as leather. Figure 6.30 shows the principle of osmosis. The pure solvent $n_1$ attempts to enter and dilute the solution $n_2$ through a semipermeable membrane, to equalize the solutions, whereby an osmotic pressure p develops against the walls. The magnitude corresponds to the height of the meniscus multiplied by the density of the solution. Van't Hoff's equation was modified by Mahan (1964) to yield:

$$P\frac{V}{n_1} = RT\frac{n_2}{n_1},$$

where R is the gas constant (0.0821 per mole-degree), T is temperature (K), V is the volume, $n_1$ is the pure solvent, water, and $n_2$ is the number of moles of solute.

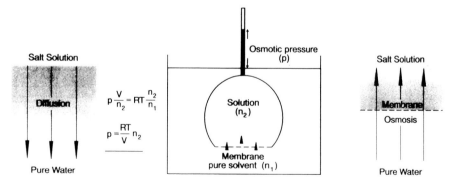

**Fig. 6.30.** Principle of osmosis and diffusion of ions; osmosis through a membrane, with the principle and calculation of pressures. (After Mahan 1964)

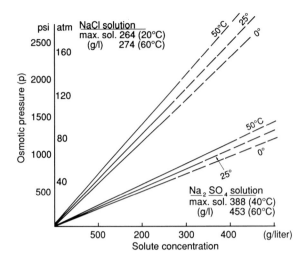

**Fig. 6.31.** Osmotic pressures of an NaCl and an Na$_2$SO$_4$ solution as a function of the solute concentration, temperature, and the osmotic pressures. Calculations employ formula given in Fig. 6.30

The plot of NaCl and Na$_2$SO$_4$ in Fig. 6.31 gives expected osmotic pressures, based on the solute concentration of the salts and on the temperature of the rock substance. Theoretical pressures may be read for various temperatures and solute concentrations for unsaturated concentrations. The process progresses in narrow capillaries along oriented water molecules remote from observation. Salt molecules can also serve as membranes. The diagram offers a theoretical evaluation of the greatest possible pressure developed by osmosis, by changes of temperatures, mole concentrations of the pore water, and the character of the membrane.

## 6.13 Honeycomb Weathering

The formation of case-hardened surfaces by the movement of dissolved ions toward the stone surface may form a temporary protective crust. Cavities of different size and shape often pockmark buildings and natural rock outcrops. The literature on this conspicuous feature is ample; the explanation of the origin of cavties, however, is controversial. Mustoe (1982) summarizes the present state of knowledge, ascribing the origin mostly to salt action, which this writer disputes.

Cavernous and honeycomb weathering evolve from a combination of surface hardening and mechanical disaggregation, often controlled by the presence of salts and by the rock fabric and structure. Winkler (1979) shows the relationship of case-hardened protective "shields", which are present on

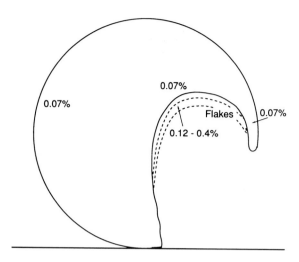

**Fig. 6.32.** Cavernous decay. Soluble salts have concentrated in the flakes and in the corroded cavity protected from rain, in granite. (After Winkler 1979)

both cavernous and honeycomb surfaces (Fig. 6.32). A careful study of honeycombs in the field and on buildings over several decades has led to the following conclusions, presented in sequence:

1. Porosity: Some stone porosity is essential to permit the travel of moisture, dissolved mineral matter, and salts. Sandstones, porous volcanic rocks, and some granites with a water sorption of 0.2 to 0.4% permit moderate to good moisture travel.

2. Movement of calcite and silica: Dissolved calcite and silica, derived from quartz and feldspars in sandstones, granites, and volcanic tuffs, move to the stone surface. The deposition of irregular layers of amorphous silica beneath the stone surface, attracted by the sun and drying winds, can form a hard layer. Calcite, available from sand grains and grain cement, goes into solution, moving into the porous sandstone and to the stone surface. The solar position is important for the movement of both silica and calcite. The uneven redeposition near the surface results in a differentially weathered surface, developing into honeycombs. Considerable amounts of salts tend to concentrate underneath a gradually eroding cavity, where they are protected from washing rain. Honeycombs in Indiana-type limestone are traced to moisture travel routes through the stone (Figs. 6.33 & 6.34). Sandstones are sometimes pockmarked with honeycombs, where moisture and the sun's action combine (Fig. 6.35). A fine example appears to be in the Elbe-Sandstein mountains of southern Saxony, on walls facing the sun in nearly calcite-free pervious massive sandstone which forms steep cliffs. With no trace of efflorescence,

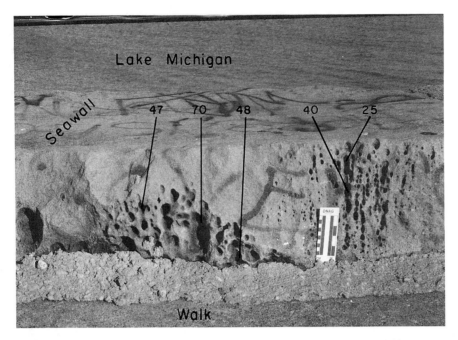

**Fig. 6.33.** Rows of honeycombs in Indiana limestone. Freshwater from Lake Michigan seeps across pervious layers during wave action. The top of the seawall is about 8 ft above the mean lake surface. Depth (mm) of the holes are given from the stone surface. North Shore at Irving Park Road, Chicago

numerous honeycombs have developed at the outer shell, which tends to separate from the rock pillars. Case hardening is here combined with the formation of honeycombs (Fig. 6.36).

3. Soluble salts: Water-soluble salts are instrumental, causing mechanical disaggregation or salt fretting in protected areas. There is evidence that the presence of salts is not required to form honeycombs; however, they can accelerate the process. Fig. 6.36 shows honeycombs in gray Ohio sandstone where deicing salts climb to 5 m above the church entrance. Honeycombs have also developed in Indiana-type limestone along the bedding planes with uneven limited permeability in a seawall that is 8 ft above the nonsaline Lake Michigan. There is no evidence of efflorescence or deicing salting in the vicinity (Fig. 6.33).

4. The introduction of calcite or gypsum from mortar, cement, or concrete into porous sandstone provides material for a partial shield. The honeycombs in quartz-rich Buntsandstein beneath a thin white crust of calcite leached from mortar joints by leakage of rainwater from the roof or condensation from inside the Münster of Freiburg suggests differential secondary cementation (Fig. 6.37). There is no calcite in the sandstone nor evidence of soluble salts.

**Fig. 6.34.** Strong differential weathering in clay-rich Molasse sand-stone. Honeycombs have developed along some bedding planes and crossbeds. The available soft stone was popular in architecture about 500 years ago, north of the Pyrenees Mountains in southern France, southern Switzerland and Germany, with a poor record of durability. Some recent replacements with chalky, gray limestone from the area started to develop honeycombs in and behind case-hardened crusts which started to flake. Cathedral of Bayonne, France

### 6.13.1 Sequence of Honeycomb Formation

1. Dissolution and differential uneven case hardening in porous rocks.
2. Small cavities begin to form along thinly hardened parts of the shield.
3. Small cavities form, aided by stone structures, loose small pebbles, and also along old tool markings on stone surfaces.
4. The presence of salts enhances crumbling inside cavities, but is not essential for the formation of honeycombs.

**Fig. 6.35.** Honeycomb surface on sandstein bluff, developed during the process of case hardening. Spalling of shells parallel to the rock surface is in evidence over almost the entire picture. Bastei, Elbe-Sandstein mountains near Bad Schandau, Saxony

## 6.14 Monuments from the Desert

Arid weathering has affected many monuments in the ancient centers of civilizations of the Near and Middle East (Knetsch 1960). Statues and monuments have been transported from the Egyptian, Iranian, Jordanian, and other desert areas of the world to museums in the more humid western cities. Rapid decay, spalling, and obliteration of inscriptions has started

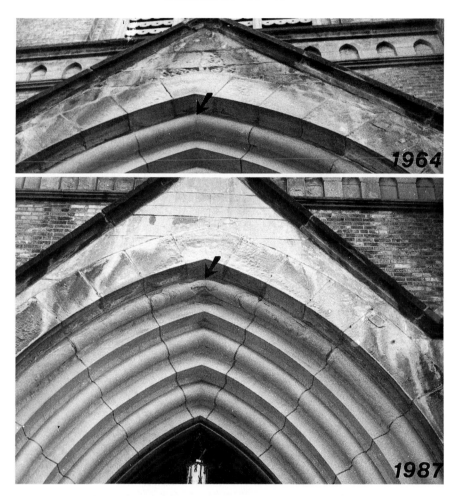

**Fig. 6.36.** Time-lapse photos of the south entrance to Sacred Heart Church, Notre Dame University, Indiana. *1964* Efflorescence beneath protected stone ledges, honeycombs above. Moisture with salts is pulled up 5 m from street level. The porous Berea quartz sandstone is free of calcite; calcite, however, was leached from the mortar joints, case hardening the stone surface. Honeycombs developed on the case-hardened surfaces (similar to Fig. 6.37). *1987* Honeycombs were covered with Portland cement (1965) and the joints replaced with putty. Salt-laden moisture could now only escape to beneath the protruding ledge, destroying the stone near the apex of the arch (*arrow*)

soon after their relocation. Museum curators soon discovered that such monuments must be leached of entrapped salts, inherited from the deserts, to enable their survival in museums in the West. The severe damage inflicted on Cleopatra's Needle in Central Park, New York, was erroneously blamed on acid rain and air pollution in the past. Analysis of the damage to the inscription challenged Winkler (1980) to link the history of the monument

**Fig. 6.37.** Development of honeycombs underneath a thin cover of a calcitic crust. Calcite was leached from lime mortar by condensation moisture near the window. The color contrast shows well against the dark red Buntsandstein with visible crossbedding and flow structure. East face of cathedral at Freiburg, Germany

with the historical implications; the emplacement of salts may have occurred in the following historical sequence:

1. 1500 to 500 B.C. Thotmes III quarried granite obelisks in Aswan and shipped them down the Nile to the ancient capital of Heliopolis, near present-day Cairo. We do not know whether salt has entered during this time, though the obelisks stood erect on well-fitted pedestals (Fig. 6.38). Annual precipitation of 3 in. is not believed to have infiltrated the stone with moisture.

500 B.C. — 16 B.C.

Salt concentration on
upper stone surface

Flood plain silts of Nile Delta
infiltrated with salts from
desert (mostly hydrateable
sulfates of Ca Mg)

**Fig. 6.38.** Sequence of historical events in the weathering of the Egyptian granite obelisk in New York City, called Cleopatra's Needle. (Winkler 1980)

2. 500 to 16 B.C. All obelisks at and near Heliopolis were overturned by the Persian king Kambyses in 500 B.C. Continuous irrigation of the Nile floodplains since 3000 B.C. has concentrated great quantities of salts by excessive recycling of irrigation water. The silts on which the micro-cracked Aswan granites lay permitted the entry and concentration of hydratable salts of Na and Mg, pulled by the sun to the upper surface of the obelisk. The entrapped salts could not exert enough pressure by crystallization or heat expansion to inflict damage on the stone (Fig. 6.38).

3. 16 B.C.–1800 A.D. The Roman emperor Augustus transported some of the toppled obelisks to Alexandria in 16 B.C., where they were erected near the Mediterranean shore, exposed to oceanic salt spray and sand drift. Signs of burial of the lower part of the obelisks in drift sand is still in evidence on both monuments, the one erected in Central Park, New York (Figs. 6.39 & 6.40), and the other on the shore of the river Thames in London in 1879.

4. 1800 to present. Scaling of the New York needle started after a few years of exposure to the humid atmosphere of New York City. A total of 700 lb of scalings fell to the ground in only two spring seasons following erection. Hot paraffin applied to the needle in 1885 stopped the process of decay (Burgess and Schaffer 1952). Hydration of the entrapped salts caused the flaking on the present east face, grading to perfect preservation on the west face, the side which pointed upward when the needle lay on the silt in Egypt. Though most of the salt had been washed out by rainwater, salt trapped deep within the stone has remained in place. Hydration was the cause for all the damage, since the action of crystallization would have attacked the stone long ago. The London needle was coated with hot wax shortly after its arrival; thus, major damage was prevented by excluding the moisture in the atmosphere. The sequence of events is illustrated in Fig. 6.38 and 6.39. All stone monuments derived from desert areas should be leached in distilled water after arrival in a museum, or sealed immediately to prevent atmospheric mosisture having access.

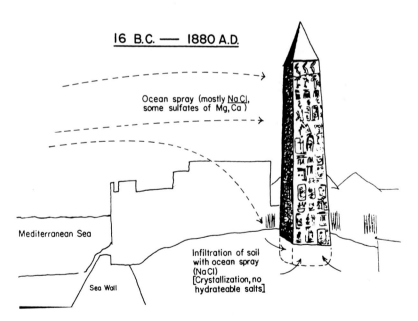

**Fig. 6.39.** *Above* Cleopatra's Needle in its setting in Alexandria (photo 1878). *Below* influx of sea salts from the Mediterranean. Historic photo from ARAMCO World (1975), sketch from Winkler (1980)

**Fig. 6.40.** Cleopatra's Needle in Central Park, New York. The inscription is sharp on the present east face, while the south and west faces have been almost obliterated. Figure 6.38 explains the process of salt infiltration which occurred while the monument was resting on its side on the flood-plain silt of the Nile for 500 years. The corrosion at the base occurred while sand of the Mediterranean shore covered the base for 1900 years. Photo taken 1982

# References

Arnold A, Zehnder K (1990) Salt weathering on monuments. Advanced Workshop. Analytical methodologies for the investigation of damaged stones. Pavia, Italy, Sept 14–21, 1990, organized by F Veniale and U Zezza, 58 pp

Braitsch O (1962) Entstehung und Stoffbestand der Salzlagerstätten. Springer Berlin Göttingen Heidelberg, 232 pp

Burgess G, Schaffer RJ (1952) Cleopatra's Needle. Chem Industry 1952:1026–1029

Cooke CW, Smalley IJ (1968) Salt weathering in deserts. Nature 220:1226–1227

Correns CW (1949) Growth and dissolution of crystals under linear pressure. Discussions of the Faraday Soc 5:267–271

Dunn JR, Hudec PP (1966) Water, clay and rock soundness. Ohio J Sci 66(2):153–168

Evans IS (1969/70) Salt crystallization and rock weathering: A review. Review de Geomorphologie Dynamique XIX(4):153–177

Fitzner B (1988) Porosity properties of naturally or artificially weathered sandstones. Proc VIth internatl Congr on Deterioration and Conservation of Stone, Torun, 12–14 Sept, 1988, pp 236–245

Kaiser E (1929) Über eine Grundfrage der natürlichen Verwitterung und der chemischen Verwitterung der Bausteine in Vergleich mit der in der Natur. Chemie der Erde IV:291–342

Kessler DW, Hockman A (1950) Thermal and moisture expansion studies of some domestic granites. US Natl Bur Stand Res Pap 2087, 44:395–410

Kieslinger A (1957) Feuchtigkeitsschaeden an Bauwerken. Zement und Beton 9:1–9

Kieslinger A (1959) Rahmenverwitterung. Geologie und Bauwesen 24(3–4):171–186

Klieger P (1980) Durability studies at the Portland Cement Association. In: Sereda PJ, Litvan GG (eds) Durability of building materials and components, ASTM, STP 691:282–300

Knetsch G (1960) Über aride Verwitterung unter besonderer Berücksichtigung natürlicher und künstlicher Wände in Aegypten. Z Geomorphol Suppl I:190–205

Kuenzel H (1988) Mechanismen der Steinschädigung bei Krustenbildung. Bautenschutz Bausanierung (Zürich) 11(2):61–68

Kwaad FJPM (1970) Experiments in the granular disintegration of granite by salt action. Univ Amsterdam, Fysisch Geografischen Bodemkundig Lab From Field to Lab Publ 16:67–80

Mahan BH (1964) Elementary chemical thermodynamics. Benjamin, Amsterdam, 155 pp

Mamillan M (1968) L'alteration et la preservation de pierres dans les monuments historiques. Etude de l'Alteration des Pierres, vol I. Colloq tenusd a Bruxelles le Ferr 1966–1967, pp 65–98, Cons Int des Monuments et des Sites, ICOMOS

Mamillan M (1981) Connaissances actuelles des problemes de remontees d'eau par capillarite dans les murs. The conservation of stone II. Cent par la Conservazione delle Sculture all Aperto, Bologna 27–30 Oct 1981, pp 59–72

Mortensen H (1933) Die Salzsprengung und ihre Bedeutung für die regional klimatische Gliederung der Wüsten. Petermann's Geogr Mitt, pp 133–135

Mullin JW (1961) Crystallization. Butterworths, London, 268 pp

Mustoe GE (1982) The origin of honeycomb weathering. Geol Soc Am Bull 93:108–115

Puehringer J (1983) Salt disintegration, salt migration and degradation by salt – a hypothesis. Swed Counc Building Res D15, Stockholm, 159 pp

Roesli A, Harnik AB (1980) Improving the durability of concrete freezing and deicing salts. In: Sereda PJ, Litvan GG (eds) Durability of building materials and components. ASTM STP 691:464–473

Schmidt-Thomsen K (1969) Zum Problem der Steinzerstörung und Konservierung. Dtsch Kunst und Denkmalpflege, pp 11–23

Snethlage R (1984) Steinkonservierung 1979–1983. Ber für die Denkmalpflege, Arbeitshefte des Laudesamtes für Deukmalpflege Arbeitsheft 22, 203 pp

Snethlage R, Hoffmann D, Knoefel D (1986) Simulation der Verwitterung von Naturstein, Teil 2: Physikalisch-chemische Verwitterungsreaktionen. In: Wittmann FH (ed) Materials science and restoration. 2nd Int Colloq, Esslingen, Sept 2–4, Stiftung Volkswagenwerk, Arbeitsheft 22, Bayer Landesamt für Deukmalpflege, pp 11–17

US Weather Bureau (1955) Pseudo-adiabatic chart, WB From 770–9. US Dept of Commerce

Vos BH, Tammes E (1969) Flow of water in the liquid phase. Rep No B 1-68-38, Inst TNO for building materials and building structures, Delft, Holland, 45 pp

Wellman HW, Wilson AT (1968) Salt weathering or fretting. In: Fairbridge EW (ed) Encyclopedia of geomorphology, vol III. Reinhold, New York, pp 968–970

Wendler E, Klemm DD, Snethlage R (1990) Contour Scaling on building facades – dependence on stone type and environmental conditions. Advanced Workshop: Analytical methodologies for the investigation of damaged stones, Pavia, Italy, 14–21 Sept 1990, 7 pp

Winkler EM (1973) Stone: Properties, durability in man's environment, 1st edn. Springer Berlin Heidelberg New York, 230 pp

Winkler EM (1977) The decay of building stones: A literature review. Bull Assoc Preserv Technol IX(4) cover and p 52

Winkler EM (1979) Role of salts in development of granitic tafoni, South Australia: A discussion. J Geol 87:199–120

Winkler EM (1980) Historical implications in the complexity of destructive salt weahtering – Cleopatra's Needle, New York. Bull Assoc Preserv Technol XII(2):94–102

Winkler EM, Wilhelm EJ (1970) Saltburst by hydration pressures in architectural stone in urban atmosphere. Geol Soc Am Bull 81(2):567–572

Winkler EM, Singer PC (1972) Crystallization pressure of salts in stone and concrete. Geol Soc Am Bull 83(11):3509–3514

# 7 Chemical Weathering

## 7.1 Introduction

The decay of rock and stone has been well known to architects and conservators of historical buildings and monuments since the days of the Roman architect Vitruvius. We understand most phases of physical and chemical weathering as the result of minerals and rocks attempting to reach equilibrium under conditions at the earth's surface. Much less is known, however, of the time required for these processes to take place in the complex environment of the earth's surface. The following summarizes the present knowledge of weathering processes and rates.

A first quantitative approach of weathering was attempted by Sir Archibald Geikie in 1880, in his study of the weathering rates of headstones in Edinburgh graveyards; this was followed by a smililar study in New York City by Julien (1884). Extensive damage to buildings and monuments during World War II exposed for study the weathered margins beneath soot-covered surfaces. The natural heterogeneity of most rocks makes the modeling of weathering rates very difficult. Granites and basalts from different locations can vary in mineral composition and physical properties. The variation of sedimentary rocks, especially sandstones, is made even greater by the near unlimited range of deposition; there may also be variations in mineral composition and physical properties. The variety of sandstones is even greater, caused by the great range of deposition and degree and kind of natural cements. No attempt should be made to predict the life of a building subjected to stone decay before one has a valid "corrosiveness index" for a given rock in the type of environment where the building will be located. We should distinguish between true solution, or dissolution, and the dissociation or solubilization of silicate minerals by hydrolysis, resulting in a stable residue, like clay.

## 7.2 Weathering by Dissolution

Dissolution is the dissociation of a mineral in a solvent, such as water. Solubility is the concentration of ions in a saturated solution of a soluble salt. It is measured in moles per liter. The amount of salts that can be chemically dissociated in pure water depends on the space available for oppositely charged ions. The solubility data are here limited to the common rock-forming minerals, calcite, dolomite, gypsum and silica, quartz, and amorphous opal (see Table 7.2). The dissolution of rock constituents is inversely proportional to the saturation of the solution and stops when saturation is reached. The dissolution rate depends on the solubility of the salt, on the degree of initial saturation of the solvent, and on the solvent movement, which can keep the solvent in contact with the stone undersaturated. Permanent undersaturation commonly occurs at the contact of stone with rainwater and stream water, on bridges or on buildings, and on jetties just below the water level.

### 7.2.1 Solvents and Solvent Motion

The following natural aqueous solvents are considered: acid rain, stream water, lake water, seawater, and groundwater. Table 7.1 summarizes the distribution of all the water available on earth.

The relatively large amount of atmospheric water is of great importance for the decay of stone.

Solvent motion is a very important factor in the dissolution of minerals and rocks. Kaye (1957) determined experimentally the influence of solvent motion on the acceleration of the solubility of acid rain and other active solvents. He used dilute HCl on marble. An exponential increase of solvent flow results in only a linear widening of the solution channel (Fig. 7.1). A flow velocity of $2\,cm^3/s$ widened the channel diameter 0.4mm, whereas a flow velocity of $10\,cm^3/s$ increased the diameter about 1.0mm, and a flow of $40\,cm^3/s$ increased it only 1.6mm. Such values may well apply to natural conditions where the doubling of the flow velocity of corrosive river water in contact with jetty stone fortunately does not double the surface corrosion.

The solution of the mineral substance, e.g., gypsum and quartz, is influenced either by the temperature, or by the temperature and pH of the solvent (Fig. 7.2), or like quartz (Fig. 7.3), and the carbonates, the

**Table 7.1.** Distribution of world water supply

| | |
|---|---|
| World oceans | 96.5% |
| Ice caps and glaciers | 1.99% |
| Groundwater | 0.617% |
| Atmosphere | 0.291% |
| Freshwater lakes | 0.071% |
| Saline lakes, inland seas | 0.048% |
| Soil moisture | 0.036% |
| Rivers | 0.014% |

Data from the US Geolgical Survey.

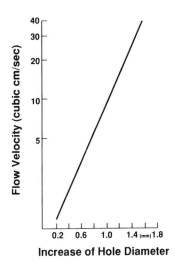

**Fig. 7.1.** The influence of the flow rate on the rate of solution of limestone by dilute HCl. (Kaye 1957)

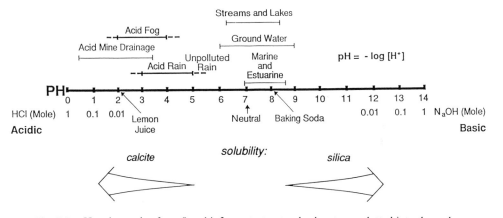

**Fig. 7.2.** pH scale ranging from 0 to 14. Important natural solvents are plotted into the scale

dependence on the temperature, the $CO_2$ and the pH content of the solvent.

## 7.2.2 Solubility by Temperature Only

Most salts increase their solubility with increasing temperature as more space becomes available for the water molecule to attach itself along dissociated ions. The solubility of gypsum is given for various temperatures in

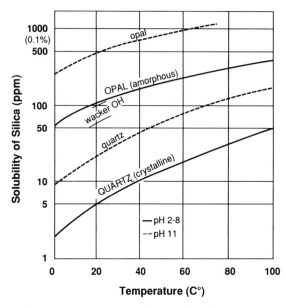

**Fig. 7.3.** Solubility of crystalline quartz, amorphous opal, and also silica gel precipitated from sandstone strengthener, Wacker OH. Solubility data adapted from Krauskopf (1956). Wacker OH values from Snethlage (1984)

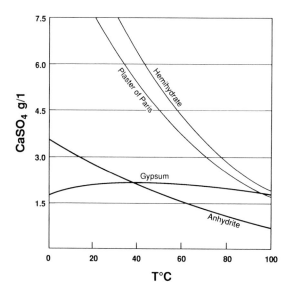

**Fig. 7.4.** Solubilities of the sulfates of calcium as a function of temperature. (Adapted from Braitsch 1962)

**Table 7.2.** Solubilities of some common important water-soluble minerals for freshwater and seawater

| Chemical compound | Solubility (ppm) | |
|---|---|---|
| | Freshwater (20 °C) | Seawater |
| Oxides | *pH 1–8 (pH 11)* | *pH 8.2* |
| opal | 105 (490) | 4.5 |
| quartz | 5 (30) | |
| Carbonates | | |
| $CaCO_3$ (calcite) | 40–85 | 66 |
| $Ca,Mg(CO_3)$ (dolomite) | 21 | 50 |
| $FeCO_3$ (siderite) | 10–25 | – |
| Sulfates | | |
| $CaSO_4 \cdot 2H_2O$ (gypsum) | 2400 | 6000 |
| $CaSO_4$ (anhydrite) | 2000 | |
| $MgSO_4 \cdot 7H_2O$ (epsomite) | 262 g/l (20 °C) | |
| $MgSO_4 \cdot 6H_2O$ (hexahydrite) | 308 g/l (20 °C) | |
| $NaSO_4$ (thenardite) | 388 g/l (40 °C) | |
| Chlorides | | |
| $MgCl_2 \cdot 6H_2O$ (bischofite) | 2635 g/l (20 °C) | |
| $NaCl$ (halite) | 264 g/l | |
| $CaCl_2$ ("ice-foe") | 730 g/l | |

Fig. 7.4. Surprisingly, the maximum solubility lies near 40 °C with almost 2100 ppm (2.1 g/l), but only 1600 ppm at near freezing, and 1575 ppm at boiling point. The true actual solubility is over 20 times greater than for that of calcite. The dissolution of gypsum can introduce large quantitites of the sulfate into water. A high sulfate content in natural waters tends to selectively attack Portland cement in concrete, leading to rapid crumbling of the concrete. Sulfate-rich waters may reintroduce gypsum by replacement of calcite into porous stone, causing efflorescence and salt fretting (Fig. 7.4). Table 7.2 summarizes the solubilities of common oxides, carbonates, sulfates, and chlorides, including the deicing salt $CaCl_2$, which readily enters stone.

The presence of the same ion tends to decrease the solubility, e.g., Ca and gypsum; gypsum dissolves a maximum of 2 g/l on limestone or marble, while the presence of other ions, like Na or Mg, in sea walls can increase the solubility threefold. The acidity of natural solvents is plotted in Fig. 7.2.

### 7.2.3 Calcite, Dolomite

The presence of $CO_2$ in water, mostly as carbonic acid, accelerates the solution process, because solid calcite may either dissolve directly or be converted to the very soluble calcium bicarbonate, as follows:

1. $CaCO_3$ (solid to $CaCO_3$ solution)
2. $H_2O + CO_2$ to $H_2CO_3$ to $Ca(HCO_3)_2$ (calcium bicarbonate).

Thrailkill's (1968) simple double logarithmic plot shows the relationship of calcite dissolution as a function of the $CO_2$ content of the solvent, the water temperature, and the pH of the water (see Fig. 7.9). The pH of the solvent, the acidity, reflects the $CO_2$ content as well as the presence of other acids. Lines of saturation are drawn for 0, 20, and 40 °C; these connect all the points at which the solvent is in equilibrium with calcite when solution stops and precipitation of calcite starts in all natural waters as the formation of dripstones, tufa, and travertine.

### 7.2.4 Quartz

The crystalline form of $SiO_2$ and the more soluble, amorphous or noncrystalline $SiO_2$, opal, dissolve well in pure water as a function of the temperature

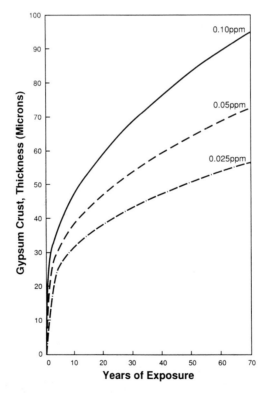

**Fig. 7.5.** Prediction for the development of a gypsum crust, as a function of the years of exposure to $SO_2$ in the atmosphere, at 0.025, 0.05, and 0.1 ppm $SO_2$, at 60 to 80% RH. (Gauri et al. 1989)

in the pH range 2–8; the solubility increases rapidly at pH 11 and above. pH values near 11 do occur in some alkaline lakes, small desert puddles, and perhaps locally in urban masonry walls (Fig. 7.2). Quartz sandstones consist of crystalline sandgrains and a usually finer grained to amorphous silica cement; this results in differential solubility that can lead to crumbling of the host rock and migration of silica in solution toward the surface, leading to case hardening and to the formation of thin surface flakes and shells. Silica

**Fig. 7.6.** Crust of soot and gypsum on Indiana limestone. Note small pustules on carved surfaces. From east face of Museum of Science and Industry, Chicago, Illinois

introduced through organic silanes will dissolve in masonry heated by the sun, often in only a few years (Fig. 7.3).

### 7.2.5 Gypsum

$CaSO_4 \cdot 2H_2O$ is a secondary mineral formed from calcite in the presence of sulfate in the atmospheric environment. Four different marbles with different grain sizes and degree of interlocking grains were tested in the laboratory for the reaction (rate constant) with sulfate; gypsum formation, however, depends on the different degree of diffusivity of the marbles, which depends on the degree of the interlocking calcite grains. The gypsum crust was thickest on Carrara marble with the least interlock of grains (Kulshreshtha et al. 1989). Figures 7.4 and 7.5 plot the thickness of the gypsum crust against the atmospheric $SO_2$ and the time of exposure, averaged from the diffusivity values (Gauri et al. 1989). Figure 7.6 shows a black crust on Indiana limestone, a mixture of gypsum and soot underneath a narrow ledge of unaffected stone. Rain-washed marble surfaces rarely accumulate gypsum. Gypsum dissolves 25 times more than calcite (80 versus 2000 ppm). It should be removable with warm water.

**Fig. 7.7.** Surface of rib of marble column against an unweathered band of hornblende crossing the rib. From the south entrance, exposed to weather, of the Field Museum of Natural History, Chicago, Illinois. (Winkler 1987)

### 7.2.6 Solubility Rates

A number of techniques have been developed to determine the reduction of stone surfaces.

1. Differential erosion of minerals with different rates of weathering on flat, machined surfaces, measured with a needle-point depth micrometer. Surface reductions can be measured from unweathered reference points, like patches or bands of hornblende, against calcite, or some dense fossil shells in limestones. Veins of quartz, pyrite, and hornblende have served as reference points in measurements made on the marble columns and horizontal balustrades of the Field Museum of Natural History in Chicago (Winkler 1987; Figs. 7.7 & 7.8).

2. Plugs and cemented inserts: Plugs of lead and inserts were often used to lift blocks of architectural stone into place in the past. Trudgill et al. (1989) describe the method used at London's St. Paul's Cathedral. Filling the lift support holes with lead plugs permitted the measurement of the surface reduction relative to the lead plug surfaces. The annual rate of degradation of the sound Portland stone averages 0.06 mm for each 1000 mm

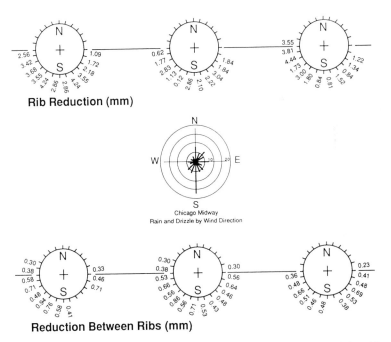

**Fig. 7.8.** Reduction of marble ribs measured against unweathered grains of hornblende at different parts of the columns. Values were measured with a micrometer, both on top of the ribs and between the ribs. South side of Field Museum of Natural History, Chicago, Illinois. (Winkler 1987)

of rainfall. Plugs are not a part of the stone. We do not yet know whether frost, pressure by osmosis, or ordered water forces out these plugs. High and Hanna (1970) permanently cemented a small tripod onto a limestone surface to measure the surface reduction in the field with an attached micrometer; values averaged 0.06 mm/year for each 1000 mm of rainfall. The method was also used on London's St. Paul's Cathedral; the data correlate with the surface reduction relative to old lead plugs.

3. Depth of inscription on tombstones: Several researchers have used tombstones to record the depth of inscription and the reduction rate of the original stone surface. The accuracy of such techniques should be questioned (Cann 1974).

4. Caliper measurements of marble tombstones: Livingston and Baer (1988) measured thousands of different marble tombstones on military cemeteries in the USA; they used large calipers to compare the supposedly unweathered bottom part of the stone against the surface reduction of the upper portion of the tombstones. Expansion by the release of locked-in stresses was not considered in their extensive study. No weathering rates were published.

5. Natural exposure tests: The controlled exposure of slabs of marble and limestone to the natural environment has been carried out by several investigators at various locations in the world. The most thorough test series was set up by the US Geological Survey (Reddy 1988) in a variety of climates. The runoff from the stone slabs was carefully analyzed and the chemical load of the runoff correlated with the surface reduction, given in $\mu$m per year. This is often given as the reduction measured on field exposures of plates of metal (see Fig. 7.10) (Sect. 5.6).

Most authors fail to give the amount and pH of precipitation; therefore, a true correlation is difficult (Table 7.3).

In Table 7.3 surface reduction of carbonates is shown per 100 years, and per 1000 mm of rainfall at a pH near 4.6. The latter presentation is preferred by this author to enable correlation with other stations.

The saturation for rainwater in rural, urban, and urban-industrial environments is plotted in Fig. 7.9. The pH of rainwater is very difficult to measure but is normally near that of distilled water. Published pH data for rainwater depend on some determining variables, like the time of removal of the samples from collecting pans after a prolonged rainfall, the length of time rainwater stands in the collecting pan before analysis, the degree of reaction that has taken place with buffering dusts in the pan, and the technique of pH analysis. pH data of rainwater should therefore be accepted with great caution (Fig. 7.10).

The rainwater with rural, urban, or industrial $CO_2$ levels moves with increasing saturation from the baseline of Fig. 7.9 toward the line of saturation, while the rate of dissolution decreases. Rural rainwater can dissolve 95 ppm of calcite at 0 °C, 120 ppm in an urban atmosphere, and 140 ppm in industrial-urban environs. The distribution of the global rain pH is given in

**Table 7.3.** Surface reduction at various locations

| Reference | Place | Surface reduction | |
|---|---|---|---|
| | | mm/100 years | mm/1000 mm rain |
| Winkler (1966) | South Bend, IN | | Rural 1 |
| Trudgill et al. (1989) | London | 8.5 | Urban 3 |
| Reddy (1988) | | | Limestone 8.07 |
| | | | Marble 6.09 |
| Dragovich (1987) | Sydney | 0.82 | |
| Attewell-Taylor (1990) | Durham | 0.2–0.9 suburban | |
| | | (1.4) industrial | |
| Feddema-Meierding (1987) | Philadelphia | 3.5 urban | |
| | | 0.5 rural | |
| Janes-Cooke (1987) | SE England | 2.9 urban London | 4.46 |
| | | 0.9 rural Garston | |
| Winkler (1990) | Chicago, marble | 2.6 horizontal | 0.016 |
| | | 2.1 vertical | 0.011 |

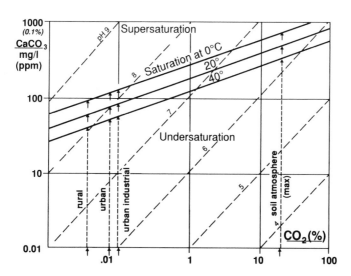

**Fig. 7.9.** Saturation – solubility graph for calcite in rural, urban, urban-industrial, and soil environments, as a function of $CO_2$ content, temperature, and the pH of the water. (Adapted from Thrailkill 1968)

Fig. 5.20. The chemistry of rainwater is discussed in detail in Section 5.4. The effect of atmospheric $CO_2$ in rainwater is discussed in Section 5.2. Marble in caves or buried in soil may be exposed to as much as 17% $CO_2$. Such a high percentage of $CO_2$ can dissolve 700 ppm calcite. The degree of undersaturation equals the corrosiveness for carbonate rocks, provided that

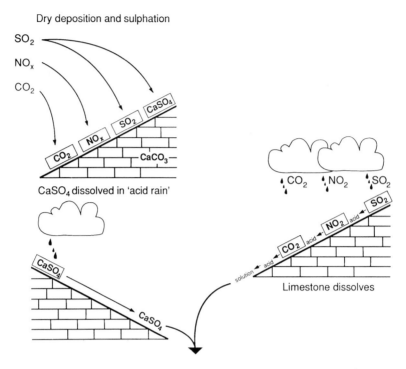

**Fig. 7.10.** Weathering of limestones and marbles in polluted atmospheres. (Adapted from Jaynes and Cooke 1987)

solvent movement does not prevent saturation. Data should be obtained from homogeneous stone materials. Calcitic fossil fragments or other components can show very different values of solubility. While fossil shells resist weathering, the looser fabric dissolves readily, leaving a measurable relief (Figs. 7.3, 7.11 & 7.12). Carbonates in a tropical humid climate record much greater solution rates than in moderately humid areas. In contrast with the solubility curves, the solubility increases with decreasing temperature. The chemical solubility appears to be superimposed by the activity of chelating microfloras.

*Stream Water.* Stone for bridges, dams, locks, and bank fortifications are exposed to chemical attack, often by quickly moving waters. In addition, mechanical attack can be observed by impacting sand, silt, and clay particles in suspension. Such sources contribute to the supply of additional ions and toward a decreasing corrosiveness. Ions are also added as atmospheric inputs and through ever-present groundwater seepage into rivers and lakes. Waters low in dissolved matter, soft waters, seek equilibrium by chemical attack on the rock they contact. The softer the waters the stronger the

**Fig. 7.11.** Modeling of steps along crossbedding by rainwater and mechanical abrasion in Indiana limestone. National Christian Church, Thomas Circle, Washington DC

attack (Fig. 7.9). Streams draining arid areas are often loaded with chloride and sulfates. The effect of aggressive components is discussed in Section 5.2.

*Lake Waters.* Rainwater, dry fallout, the leaching of rock on shore lines, and groundwater influx all supply ions. Small lakes fluctuate frequently, while large lakes have a more consistent composition. Softwater lakes can have a calcite concentration of less than 40 ppm at a pH of 6.8–7.4. Hard-water lakes can draw 40–200 ppm of calcite with a pH of 8.0–8.7. Such lakes are generally lime saturated and are less corrosive.

*Seawater.* Seawater normally has a total of 3.5% of salts in solution and can be corrosive to stone and metals. Despite the general saturation of seawater with $CaCO_3$, beachrock erosion is well known in calcareous rocks used on sea walls and jetties. Revelle and Emery (1957) observed strong dissolution at the intertidal zone between high and low tides, where strong seasonal and temporary undersaturation can occur by the $CO_2$ emission during excessive algal photosynthesis. Cold seawater and very diluted brackish harbor waters may also become locally undersaturated, and thereby corrosive. Molecules of hydrocarbons floating in coastal waters can, if abundant, form resistant coatings on sandgrains and shell fragments, protecting the inorganic mineral

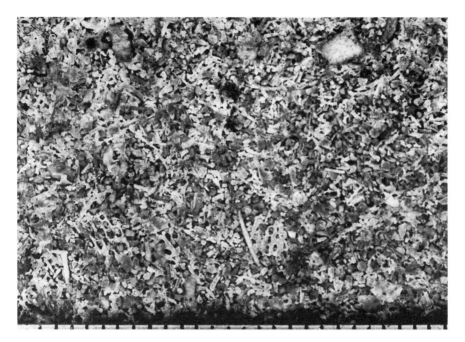

**Fig. 7.12.** Weathered surface on Indiana limestone. The shell fragments stand out against the cementing matrix, which dissolves readily. Balustrade on administration building, Notre Dame, Indiana. Scale is in mm

substance from chemical attack by seawater. Only grains with a patchy organic skin become vulnerable to solution attack. Foraging organisms can also break this protective coating.

*Groundwater.* Groundwater occasionally contacts stone and concrete in deep foundations. The slow flow of groundwater through rock pores and narrow channels allows enough time for the water to reach near equilibrium with carbonate rocks (Fig. 7.13). The ion content of groundwater is subject to strong regional variation, depending on both the source rock and the length of travel of the water. Glacial drift and limestone areas provide enough $CaCO_3$ to saturate such waters, rapidly neutralizing their corrosiveness.

## 7.3 Complex Solubility Features: Stylolites

Stylolites, or "crow's feet", form irregular wavy seams on a stone surface as the result of interstratal pressure solution in many limestones and limestone-marbles. The upper strata tend to collapse onto the lower undissolved beds,

**Fig. 7.13.** Solution channels in Indiana limestone quarry, narrowing along denser beds and widening along porous shell beds. Quarry face is smoothed by the cutters of the channeling machines. Bedford, Indiana

crushing grains and filling the serrated gaps with insoluble residual clay. Stylolites most frequently run nearly parallel with the bedding planes, but may also develop an occasional vertical network. Decay along stylolites has led to complex and rapid deepening of the relief, with little hope for repair. The thickness of the clay layer along the stylus-like seam, the insoluble residue, reflects the amount of soluble $CaCO_3$ removed by dissolution. A sudden change of stone fabric and color across such stylolites should be expected. Groshong (1988) summarizes and updates the information by Park and Schot (1968) on these often complex features. Stylolites can affect features of interest to the stone industry. For example, they can influence the fabric as:

1. Grain-to-grain stylolites with sutures developed around relatively less-soluble grain-like dense fossil shells, ooids, or quartz grains (Fig. 7.14).
2. Transgranular stylolites, with sutures developed across the grains by similar solubility. Most of the frequent sutures in Tennessee and Missouri limestone-marble run across most grains. Occasionally, the sutures are offset by vertical fault-like discontinuities by the collapse of the upper portion onto the lower undissolved part (Fig. 7.15).

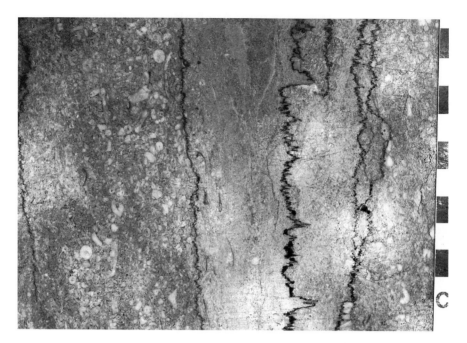

**Fig. 7.14.** Stylolite sutures, "crow's feet", separating different stone fabrics. Sutures widen upward toward a continuously wetted indoor fountain pool. Carthage, Missouri limestone-marble, Stanley Hall, Field Museum of Natural History, Chicago, Illinois. Scale is in cm

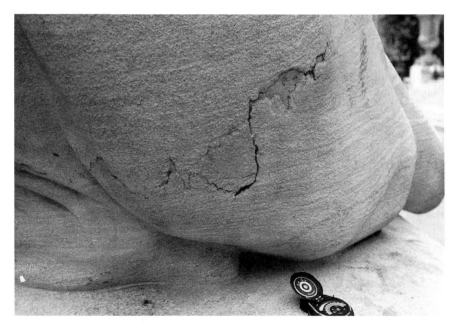

**Fig. 7.15.** Deep weathering along stylolite at the apices where the clay filling is thickest. Indiana limestone. Foot of lion statue in front of New York Public Library, New York

### 7.3.1 Strength and Durability Along Stylolites

While an even bedding plane separates the beds along more or less thin clay partings, interlocking sutures cohere well, despite the clay layers. The clay layers may range in width from thin films to layers thickening sometimes to a few millimeters at the apex (Fig. 7.16). The type of clay mineral depends on the kind of influx of mud from rivers into the ancient oceans or from land as airborne volcanic dust or soil. Mixed-layer clays and montmorillonite tend to liquify upon vibration when water soaked, e.g., during earthquakes or in heavy traffic. While no such separation was observed in laboratory tests under vibration, evidence for a potential liquification appears to be counteracted by the interlocking cog-like surfaces of most stylolites (Winkler 1989). Figure 7.17 shows a column of the temple of Apollo Epicurios at Bassi near Olympia, Greece, which is cut by numerous stylolites with clay seams weathered more than 25 mm deep. Annual precipitation of 1200 mm and frost at an elevation of 2000 m have caused deep erosion. Many earthquakes have offset the column sections against one another but have failed

**Fig. 7.16.** Deep erosion to 1 in. depth along clay-filled stylolites set vertical parallel with bedding planes. Tennessee limestone-marble. Rockne Memorial shower stall, Notre Dame, Indiana

**Fig. 7.17.** Limestone column with deeply weathered stylolites running parallel with bedding planes, not visible on photo. Earthquakes rotated the stone sections; no rotation has taken place along the wavy stylolites. Temple of Apollo Epicurios at Bassi, near Delphi, Greece. Temple was built 2400 years ago. The stylolites were mistaken for bedding planes. (Andronopoulos et al. 1988)

to rotate these along the clay-filled stylolite surfaces. Considerable surface reduction along thickened clay surfaces of the stylolites can also be observed indoors; for example, shower stalls and water fountains have shown deep weathering to below the finished surface (Fig. 7.16). The abrupt change of the lithology and rock color and the often well-accented zigzag lines give the stone character as a building and decorative material. The width of the clay seams should be closely observed. Frequent contact with running or splashing water may erode deep ugly furrows, especially when oriented vertically (Fig. 7.16).

## 7.4 Weathering of Crystalline Marble

While the weathering attack of limestone by surficial dissolution is simple, the decay of marble contrasting limestone is a complex interaction of several processes, triggered by dissolution attack, release of locked-in internal stresses by the anomalous temperature behavior of calcite, and by the expansive action of trapped moisture. Dilation can occur, resulting in the crumbling of the material in a relatively short period of time. The visible features of marble decay are surface sanding, crumbling and cracking of ribs, edges, and ledges, and bowing (scalloping) of thin vertically suspended stone panels (curtain walls). Winkler (1988) summarized the complex process of marble decay.

1. Surface reduction: The mechanism of dissolution attack is the same for all carbonate rocks. The measured rate, however, differs. Veins or patches of quartz or hornblende on a flat surface make reliable reference points for the measurement of the surface reduction (Fig. 7.8). Decaying pyrite produces sulfuric acid, which tends to accelerate and deepen the corrosion of adjacent sulfide minerals. A surface reduction of about $3.5\,\mu m/1000\,mm$ for a pH of 4.6 should be expected. Minor differences in pH values do not greatly change the surface reduction. Surface reduction may also accelerate in the presence of sulfate, converting calcite readily to the much more soluble sulfate of calcium, the gypsum (Table 7.2).

Sanding on some fine-grained marble surfaces is the first indication of the effect of dissolution by circumgranular surface attack. Small grains of calcite were observed in an irregular honeycomb-like pattern, visible only under magnification. The separating ridges between shallow pits are loosely cemented grains of redeposited calcite between the grains. Capillary action attracts and lightly cements the grains. The process appears to be related to the formation of honeycombs on rock surfaces. Stereo-photographs readily disclose such shallow ridge-and-pit microtopography (see Sect. 6.13).

2. Crumbling of marble: Thin ribs and ornaments, slabs, and tombstones tend to decay rapidly, especially fine-grained varieties. Pieces of ornaments have often decayed so deeply that water-soaked fragments have fallen to the ground. Ribs and corner sections tend to expand uninhibitedly and can crack in the center of the rib or corner piece. Ribs on a marble column of the Field Museum of Natural History in Chicago offer a good example of expansion cracking (Fig. 3.19).

Evaporation, caused by the sun and drying winds, moves dissolved calcite toward the stone surface, leading to case hardening. Dissolution-case hardening combined with stress relief can create unusual patterns on a building with exterior marble (Fig. 7.18). Like in sandstones, this may lead to the development of surface shells with a weaker zone underneath.

3. Cracking is initiated by internal stresses which were locked into the rock during metamorphic processes when limestone was recrystallized to

**Fig. 7.18.** Crumbling Vermont marble. A medium-grained, case-hardened window sill, slanted outward, is underlain by a crumbling band forming a depression to 20 mm deep. Dilation of the upper ledge is by heat–cold expansion cycles and subsequent case hardening with material leached from underneath. South side of Curtis Center, Philadelphia, Pennsylvania

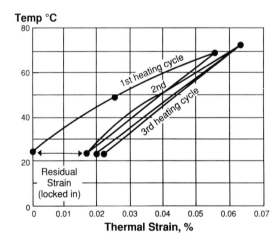

**Fig. 7.19.** Thermal strain path for a medium-grained crystalline marble. Most of the residual strain is released during the first heating cycle. (After Sage 1988)

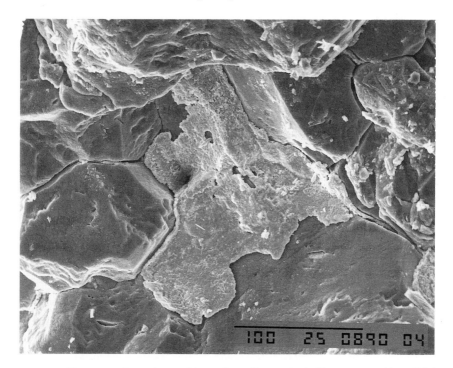

**Fig. 7.20.** Circumgranular microcracking of medium-grained Vermont marble, slightly weathered. Scanning electron micrograph, 350 × magnification. *Bar* is 100 μm long

crystalline marble hundreds of million years ago. The release of such stresses causes cracking, first along the edges where free expansion can take place around the weaker mineral contacts, and then along crystal imperfections. In contrast to limestones, most crystalline marbles crack readily and rapidly from the surface inward. Dilation of the stone fabric is accelerated where the shape of the sculpture, rib, ledge, edge, or corner permits limited expansion (Fig. 7.20).

4. Thermal cracking: When heated, calcite crystals expand along the long or c-axis, but contract along the short or b-axis. The coefficient of thermal expansion is norally 21 μm/m per °C. The coefficient of contraction is −5.4 μm/m per °C. A 5-mm calcite crystal therefore expands 4.2 μm along the long axis when heated at 40 °C, and contracts 1.08 μm along the short axis. The mechanism is pictured in Fig. 7.19, after Sage (1988). Diurnal atmospheric heating-cooling cycles of stone are believed to be sufficient to dilate marble in a matter of a few years (Chap. 5). Widening of cracks (Fig. 7.20) progresses with continued thermal cycles, resulting in an increase of the porosity and water sorption, leading to crumbling.

5. Moisture expansion: Warming water from 10 to 50 °C can exert 200 atm pressure against the capillary walls as a tensile force when the water

is trapped in narrow stone capillaries (Fig. 7.22). A high relative humidity readily fills capillaries. This is believed to be the most powerful dilating agent in stone decay. In humid climates the possible expansion effect of "ordered" water may also contribute to dilation and decay. Bowing of thin marble panels suspended on or between fixed anchors supports this explanation.

6. Spalling: Strees relief of metamorphic marble blocks can cause spalling in concentric shells, thus relieving internal stresses. Volume expansion by stress relief is discussed in Section 3.3 (Figs. 3.13–3.18). Figure 7.21 shows such spalling of fine-grained Pentelian marble on a stone block on the Acropolis of Athens.

7. Bowing of marble slabs: Facing tall buildings with thin panels of white Carrara marble has become fashionable in the last 20 years as an

**Fig. 7.21.** Surface spalling on a block of fine-grained Pentelian marble by the relief of locked-in internal stresses. Acropolis, Athens, Greece

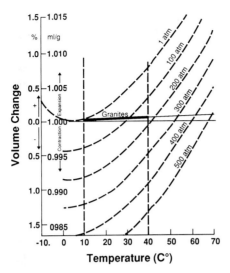

**Fig. 7.22.** Temperature – volume change – pressure of water superimposed upon the temperature expansion of granites, between 10 and 40 °C. Internal stresses of 170 atm can develop in the capillaries of granites under such conditions. Data from the International Critical Tables (1926)

expression of architectural distinction. The bowing of these marble slabs is associated with a loss of strength and they may threaten to fall off the front of the building. Unfortunately, the legendary durability of this stone is greatly reduced in humid climates and does not last longer than 10 years (Fig. 7.23). The Lincoln First Tower in Rochester, New York, was one of the first major failures of these marble slabs. The warping panels of white Carrara marble were removed from the 93-storey Amoco building in Chicago. They were replaced with panels of nearly white Mt. Airy granite from North Carolina (see Fig. 3.16). The designers have also overlooked that the Mt. Airy granite quarry floor is tenting by the extensive release of internal locked-in stresses (Chap. 3). The durability of marble cladding is summarized by Cohen and Monteiro (1991). Bowing of granite panels was also observed in the Greenwood cemetery in the subtropical humid New Orleans, Louisiana. Slabs of Georgia Cherokee marble started bowing on the Marshall Field building, on the side exposed to the sun, in Chicago after only 9 years of exposure. Warping of Carrara marble slabs, loosely attached to a backing by three or four pins, is long known from the New Orleans cemeteries in a subtropical humid climate, first described by Bucher (1956). This phenomenon is not found in arid and semiarid climates where atmospheric moisture is low. The reason for the bowing of marble slabs appears to be a combination of processes that occur in the following sequence.

**Fig. 7.23.** Bowing of thin panels of Carrara marble as cladding, exposed to the sun and moisture for only 15 years. Marble panels were replaced in 1992 with panels of Mt. Airy white granite. Amoco building, Chicago, Illinois

a) Dissolution triggers microcracking, followed by dilation from the relief of internal stresses. Without dissolution the processes of dilation and decay would probably not occur.

b) Diurnal thermal expansion–contraction cycles continue to dilate the stone, leading to expansion, bowing, and loss of strength.

c) Moisture from rain and relative humidity quickly fills pore spaces. Moisture expansion in heating–cooling cycles dilates the stone further. Bowing of stone panels is normally not observed in desert areas.

Water absorption is a good indicator of the strength and degree of disintegration of crystalline calcite marble. Quarry-fresh, dense, pure calcite marble has a water absorption of close to 0.1% and a modulus of rupture near 12 MPa (1680 psi). Winkler (1986) plotted the relationship between water absorption and the modulus of rupture (Fig. 2.11). Bowing of granite veneer panels shares some of the complex factors with marble panels, such as the action of moisture dilation, stress relief, and moisture action by microcracking by the excessive contraction of the quartzes during cooling.

Prevention of bowing: The access of moisture to marble and granites can be prevented by waterproofing unweathered stone panels. Potential discoloring of all organic chemicals exposed to UV radiation should be expected. Yellowing, often in patches, can develop on white marble surfaces in a few years.

## 7.5  Weathering of Silicate Rocks

Silicate minerals have tight crystal lattices, with $SiO_2$ tetrahedra acting as the "skeleton" to which aluminum, iron, and other metal cations (Ca, Na, K, Mg, and Fe) are attached. The tightness of such a lattice determines the ease of separation from the almost inseparable $SiO_2$ molecules. Leaching or solubilization can unlock and free some or all of these ions and fill the void spaces with water; as chemical weathering continues, clay and quartz grains form an insoluble residue, the soil. The chief ions, Ca, K, Mg, and Fe, remain in the soil attached to the surface of the clay minerals; they either become nutrients for plants or move in groundwater to streams and eventually the oceans. Vegetation binds most of these ions in various concentrations. Na, however, is rejected by all plants and is therefore readily concentrated in soils and masonry. In urban areas this natural process is further aggravated by the addition of deicing salts.

The process of silicate weathering is generally slow in human terms, but rapid in geological terms. Loughnan (1969) reports the following possible processes involved in silicate weathering: ions exposed at the surface of the lattice are unsaturated and hydrate rapidly when exposed to moisture. $H^+$ ions with their high charge tend to penetrate the mineral surfaces, readily breaking down the silicate structure (Fig. 7.24). The following processes appear to take place.

The parent mineral breaks down with the release of cations, whereby the silica in the rock may either retain its original atomic arrangement or enter the dispersed state. Singer and Navrot (1970) studied the amount of cation removal from a basalt boulder in the Negev desert, Israel:

$$Fe \gg Mg > Na > Ca > Si > K.$$

The observation forms a guideline for the degree of cation leaching. Iron is leached from biotite, mica, and hornblende first as the greenish or black soluble ferrous iron spreads across the stone surface. It readily oxidizes to an insoluble ochre-brown or yellowish ferric hydroxide. Iron can also be

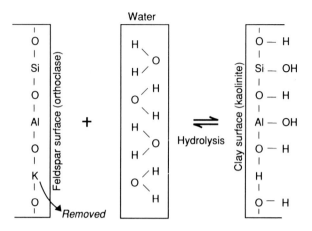

**Fig. 7.24.** Mechanism of feldspar weathering by hydrolysis: from orthoclase to kaolinite. (Loughnan 1969)

introduced as dust from railroad yards and steel mills, and from grains of pyrite. The common natural "patina", yellowing of gray granite or white marble surfaces, is an indication of initial weathering. Granites which show such a patina should be carefully tested for soundness prior to their use on exteriors. Keller et al. (1963) in their basic study tried to quantify the weathering of silicate minerals by leaching common rock-forming silicates with both distilled and $CO_2$-charged water. The latter doubled the leaching rates for Ca, Mg, Na, and K. Orthoclase feldspar and muscovite mica are much more sensitive to $CO_2$-charged waters than are other minerals. Tests performed in the laboratory by Aires-Barrios et al. (1975) using distilled water show K to be leached the most, followed by Na, and finally iron is leached the least. Leaching in a 3.5% NaCl solution almost doubles the loss of ions and increases the loss of Fe by an alarming factor of 7. Chemical cleaning of stone therefore ought to be undertaken with great caution. Occasional halos of iron rust at the outer fringe of stone corroded by deicing salt can be expected in many rocks (Sect. 6.11).

Much attention should be paid to the weathering of feldspars, which comprise nearly two-thirds of all minerals present in igneous rocks. Leaching of feldspars dulls the polish of light-reflecting surfaces and forms a thin, light, dull gray clay film. The chemical weathering of feldspars, a process of leaching and hydration, is sketched in Fig. 7.24.

Ferro-magnesian silicates, such as hornblende, hydrate to brown clay after the removal of iron. Biotite bleached by the loss of iron changes to a mixed-layer clay.

*Weathering Crusts.* Dense igneous rocks tend to weather from the surface inward; Černohouc and Šolc (1966) plotted the weathering rate of dense

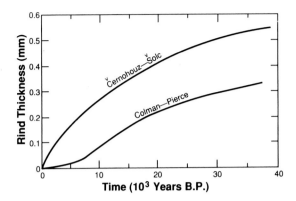

**Fig. 7.25.** Weathering rates measuring the thickness of the weathering rind in basalts. Černohouz and Šolc's (1966) curve is for basalts in Bohemia, and Colman and Pierce's (1981) curve is for basalts in the West Yellowstone area, USA

Bohemian basalts. Colman and Pierce (1981) made a similar study of basalts from the western United States. On both curves in a semihumid to humid climate the rind formation decreases with time as the weathering agents reduce accessibility to the unweathered rock surface. The slow formation of rinds in dense silicate rocks may be observed in ancient quarries (Fig. 7.25). Iron-rich margins in true granites replace near-impervious clay rinds with an adequate pore space to permit moisture travel. Iron-stained margins develop faster in sandstones.

*Degree of Weathering.* The degree of kaolinization of igneous rocks strongly determines the rock strength and durability. This is reflected by the dry-to-wet strength ratio. The rate of silicate weathering is therefore difficult to evaluate because too many factors influence the process. Strakhov (1967) estimates an increase of the weathering rate by solubilization from 2 to 2.5 times with each temperature increase of 10 °C. In most tropical areas the rate can be 20 to 40 times as fast. Such data are important guidelines for the stone industry.

*Granite Weathering.* Irfan and Dearman (1978) have observed general stages of decay of Cornish granite. These stages can be applied to all granitic rocks. Four stages are recognized:

1. Fresh granite: Mineral constituents are unweathered. Feldspars and hornblende retain their glassy luster and original colors. The freshness of feldspars can also be checked with a pocket knife to demonstrate the hardness. The fabric appears unfractured.
2. Partially stained granite: Most mineral constituents still appear sound. Plagioclases are partially gritty and turning dull; biotite is slightly decomposed. There is partial staining along widening microcracks.

3. Completely stained granite. Plagioclases are partly decomposed to soft clay aggregates. There is staining of the entire rock material, which is intensely microfractured throughout.
4. Weakened granite: Slight weathering of potash feldspars; bleached quartz is microfractured. Opening grain boundaries; intense microfracturing.

While the formation of a weathering rind will not develop in granitic rocks in the lifetime of a building, the degree of discoloring should be carefully observed as a warning for the progressive weakening of the stone.

*Weathering of Serpentine (Verde Antique).* The greenish to bright green serpentine, a hydrous magnesium silicate, is often cut by numerous white veins of calcite ($CaCO_3$) or magnesite ($MgCO_3$). It is a very stable interior decorative stone, but is readily attacked outdoors by a polluted urban atmosphere with a high $SO_4$ content. The green serpentine loses luster, color, and hardness to form a rough gray surface of talc and powdery sulfates of Ca (gypsum), or the very soluble sulfate of Mg, which readily effloresces across the exposed stone surface.

*Weathering of Basalt.* Basalt is occasionally used as dimension stone, but more often as decorative stone when found in columnar form, or as basalt slag (scoria). Resistant to weathering, huge heads carved from basalt remained buried in corrosive, tropical soil near coastal Vera Cruz, Mexico, for 4000 years and suffered only minimal surface attack (Fig. 7.26).

**Fig. 7.26.** Head of basalt. Holes and deeply weathered joints after burial in tropical soil for over 4000 years are visible. Olmec culture, Vera Cruz, Mexico. Head exhibited at hemisphere 69, San Antonio, Texas

**Fig. 7.27.** Basalt, weathered by the rapid decay of the mineral analcite, is called sun-burned basalt (Sonnenbrenner basalt). The numerous *white dots* are the clay resildue. (See Ernst 1960)

Some basalts develop light gray spots as the rock strength diminishes. This phenomenon is known as "sun-burned basalt" or Sonnenbrenner and was believed to be caused by solar radiation (Fig. 7.27). Ernst (1960) ascribes the rapid decay of the mineral analcite, a hydrous sodium aluminum silicate, as responsible for the effect. Laboratory tests, such as the immersion of pieces of basalt in KOH for 24 h, can reveal this condition before the stone is used.

# References

Aires-Barrios L, Graça RC, Veclez A (1975) Dry and wet laboratory tests and thermal fatigue of rocks. Engineering geology, vol 9. Elsevier, New York, pp 249–265

Andronopoulos B, Tzitziras A, Koukis G (1988) Engineering geological investigations in the area of Apollo Epikourios Temple at Phigalia (Peloponnesos, Greece). In: Marinos PG, Koukis GC (eds) Engineering geology of ancient works, monuments and historical sites. Athens, Sept 19–23, 1988. Balkema, Roherdam, pp 479–486

Attewell PB, Taylor D (1990) Time-dependent atmospheric degradation of building stone in a polluted environment. Environ Geol Water Sci 16(1):43–55

Braitsch O (1962) Entstehung und Stoffbestand der Salzlagerstätten. Springer, Berlin Göttingen Heidelberg, 232 pp

Bucher W (1956) Role of gravity in orogenesis. Bull Geol Soc Am 67(10):1295–1318

Cann JH (1974) A field investigation into rock weathering and soil-forming processes. J Geol Educ 22:226–230

Černohouc J, Šolc I (1966) Use of sandstone vanes and weathered basaltic crusts in absolute chronology. Nature 212(5064):806–807

Colman SM, Pierce KL (1981) Weathering rinds on andesite and basaltic stones as a quaternary age indicator, Western United States. Geol Surv Prof Pap 1210:56

Cohen M, Monteiro PJM (1991) Durability and integrity of marble cladding: a state-of-the-art review. Am Soc Civil Eng. J Performance Coustruct Facil 5(2):113–124

Dragovich D (1987) Measuring stone weathering in cities: Surface reduction on marble monuments. Environ Geol Water Sci 9(3):139–141

Ernst Th (1960) Probleme des "Sonnenbrandes" basaltischer Gesteine. Z Dtsch Geol Ges 112:178–182

Feddema JJ, Meierding TC (1987) Marble weathering and air pollution in Philadelphia. Atmos Environ 21(1):143–157

Gauri KL, Kulshreshtha NP, Punuru AR, Chowdhuri AN (1989) Rate of decay of marble in laboratory and outdoor exposure. J Mater Civil Eng 1(2):73–85

Geikie A (1880) Rock weathering as illustrated in Edinburgh church-yards. Proc R Soc Edinb 1879/80:518–532

Groshong RH (1988) Low temperature deformation mechanisms and their interpretation. Geol Soc Am Bull 100(9):1329–1360

High C, Hanna FK (1970) A method for the direct measurement of erosion of rock surfaces. Br Geomorphol Res Group Tech Bull 5:24

International Critical Tables (1926) Natl Rescarch Council, E Washburn (ed). McGraw Hill, New York

Irfan TY, Dearman WR (1978) The engineering petrography of a weathered granite in Cornwall, England. Q J Eng Geol 11:233–244

Jaynes SM, Cooke RU (1987) Stone weathering in southeast England. Atmos Environ 21(7):1601–1622

Julien AA (1884) The durability of building stones in New York City and vicinity. US 10th Census, 1880, Spec Rep on Petroleum, Coke, Building Stone 10(5):364–384

Kastner M (1981) Authigenic silicates in deep-sea sediments: Formation and diagenesis. In: Emiliani C (ed) The sea, vol 7. The oceanic lithosphere. Wiley, New York, pp 915–940

Kaye CA (1957) The effect of solvent motion on limestone in solution. J Geol 65:35–46

Keller WD, Balgoard WD, Reesman AL (1963) Dissolved products of artificially pulverized minerals and rocks. J Sediment Petrol part I, 33(1):191–204, part II, 33(2):426–437

Krauskopf BK (1956) Dissolution and precipitation of silica at low temperatures. Geochim Cosmochim Acta 10:1–26

Kulshreshtha NP, Punuru AR, Gauri KL (1989) Kinetics of reaction of $SO_2$ with marble. J Mater Civil Eng 1(2):60–71

Livingston RA, Baer NS (1988) The use of tombstones in the investigation of the deterioration of stone monuments. In: Marinos PG, Koukis GC (eds) Engineering geology of ancient works, monuments and historical sites. Athens, Sept 19–23, 1988. Balkema, Rotlerdam, pp 859–867

Loughnan F (1969) Chemical weathering of silicate minerals. Elsevier, New York, 154 pp

Park WC, Schot EH (1968) Stylolites: their nature and origin. J Sediment Petrol 38(1):175–191

Reddy MM (1988) Acid rain damage to carbonate stone: A quantitative assessment based on the aqueous geochemistry of rainfall from stone. Earth Surface Processes and Landforms 13:335–354

Revelle R, Emery KO (1957) Chemical erosion of beach rock and exposed reef rock. US Geol Surv Prof Pap 260-T:699–709

Sage ID (1988) Thermal microfracturing of marble. In: Marinos PG, Koukis GC (eds) Engineering geology of ancient works, monuments and historical sites. Athens Sept 19–23, 1988. Balkema, Rotterdam, pp 1013–1018

Singer A, Navrot J (1970) Diffusion rings in altered basalt. Chem Geol 6:31–41

Snethlage R (1984) Steinkonservierung 1979–1983. Berichte für die Stiftung Volkswagenwerk. Arbeitsheft 22, Bayer Landesamt für Denkmalpflege, Munich, 202 pp

Strakhov NM (1967) Principles of lithogenesis. Oliver and Boyd, Edinburgh, 245 pp

Thrailkill J (1968) Chemical and hydrologic factors in the excavation of limestone caves. Geol Soc Am Bull 79(1):19–46

Trudgill ST, Viles HA, Inkpen RJ, Cooke RU (1989) Remeasurement of weathering rates, St. Paul's Cathedral, London. Earth Surface Processes and Landforms 14:175–196

Winkler EM (1986) A durability index for stone. Bull Assoc Eng Geol 23(3):344–347

Winkler EM (1987) Weathering and weathering rates of natural stone. Environ Geol Water Sci 9(2):85–92

Winkler EM (1988) Weathering of crystalline marble. In: Marinos PG, Koukis GC (eds) Engineering geology of ancient works, monuments and historical sites. Athens, Sept 11–23, 1988. Balkema, Rotlerdam, pp 717–721

Winkler EM (1989) Deep weathering of stylolites in limestone on the columns of the temple of Apollo Epicurios, Greece. The conservation of monuments in the Mediterranean Basin. In: Zezza F (ed) Proc 1st Int Symp, Bari 1989, pp 185–187

# 8 Stone Decay by Plants and Animals

Plants and animals attack stone by both mechanical and chemical action. Higher plants can affect stone both mechanically and biochemically; bacteria, the lowest kind of life, only attack by chemical means. The biotic decay is very complex and not yet fully understood.

## 8.1 Bacterial Contribution to Stone Decay

Bacteria are classified by their activity as autotrophic or heterotrophic: While autotrophic bacteria take their energy from the sunlight, chemical oxidation, and reduction, heterotrophic bacteria obtain their energy from existing organic substances. Identification of the microorganisms is often impossible and their action is unclear. Sound stone is little populated by bacteria. Fungi start their activity during initial weathering, finding a foothold on other organisms (Webley et al. 1963). Weathered rock develops a large bacterial population along all surfaces and cracks. A clean rock surface has few bacteria and a longer life than rock populated with bacteria.

### 8.1.1 Autotrophic Bacteria

Autotrophic bacteria play an important role in incipient weathering of minerals and rocks, though the damage is rarely in evidence. According to their activity we distinguish and discuss here nitrogen-fixing bacteria, sulfur bacteria, iron bacteria, and calcite bacteria. The literature is ample and widely scattered often showing the uncertainty of microbial action and proper identification. While the soil bacteria have been well studied and identified, the bacteria on stone are less known.

#### 8.1.1.1 Nitrogen-Fixing Bacteria

The conversion of $CaCO_3$ to $Ca(NO_3)_2$ by bacteria was observed on limestone walls by Kauffmann (1960). The chemical process carried out by nitrifying bacteria appears to explain the presence of both the nitrate and sulfate of Ca:

$$2CaCO_3 + (NH_4)_2SO_4 + 4O_2 = Ca(NO_3)_2 + CaSO_4 + 2CO_2 + 4H_2O.$$

*Nitrobacter* and *Nitrosomonas* are the most important bacteria. Kauffmann believes that nitrification produces some gypsum on carbonate surfaces; the main bulk of gypsum, however, is formed by the direct interaction of sulfates in the atmosphere with protected carbonate surfaces. The mobilization of Ca from granite sand by nitrification was accelerated by a factor of 10. Berthelin (1983) schematizes in general terms microbial weathering of minerals and rocks under the influence of bacteria. This exchange is also important in the breakup of carbonates. Once the breakup process has started, acceleration by carbonic acid and other organic acids, produced from the organic remains, takes place. The acid $H^+$ continues its attack on the still unweathered mineral grains after a film of clay has formed on the mineral surface.

## 8.1.1.2 Sulfur Bacteria

*Thiobacillus thiooxidans* uses reduced sulfur compounds and *Thiobacillus thioparus* elementary sulfur. These are the most common sulfur bacteria converting sulfur dioxide gas in the atmosphere to acid, which attacks carbonates and converts these to the sulfate of Ca, the soft and very soluble gypsum. Millot and Cogne (1967) found several thousand *Thiobacillus* per gram of sandstone on the cathedral of Strasbourg. There may be as many as 100 000 organisms/g of rock (Pochon and Jaton 1967). Voûte (1969) reports 16 million bacteria/g of rock on the Buddhist temple of Borobudur in the rainforest of Java. Small crystals of gypsum were found between the sandgrains; these were produced by the conversion of the 2% of calcite grain cement. *Desulfofibrio desulfuricans* can form the nearly insoluble calcium sulfite ($CaSO_3$). While the polluted urban atmosphere appears to be the dominant factor in the direct conversion of calcite to gypsum, bacterial action appears to prevail in clean rural areas. Kauffman (1960) pictures the distribution of sulfate crusts and lesions on a stone wall with different profiles and permeability to moisture. The same profile was also used by Mamillan (1968) to illustrate the travel of moisture.

## 8.1.1.3 Iron Bacteria

The autotrophic *Ferrobacillus ferrooxidans* and possibly heterotrophic *Thiobacillus ferrooxidans* are the most important oxidizers of iron. Iverson (1974) gives a detailed account of the action of iron bacteria. These obtain their energy from the oxidation of ferrous ions to ferric ions. Sulfuric acid is

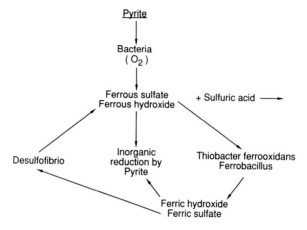

**Fig. 8.1.** Bacterial weathering of pyrite, leading to oxidation and reduction of iron and sulfur. (Modified from Silverman and Ehrlich 1964)

produced as a result of the oxidation of pyrite (Fig. 8.1). The autotrophic and heterotrophic natures are difficult to separate. Silverman and Ehrlich (1964) sketched a flow-chart of chemical reactions induced by both iron and sulfur bacteria.

### 8.1.1.4 Calcite Bacteria

*Arthrobacter* is frequently found on limestone surfaces where it participates in the decay of carbonate rocks.

### 8.1.2 Heterotrophic Bacteria

Heterotrophic bacteria live off available organic matter which has been produced by higher organisms, like lichens, on stone surfaces. Heterotrophic bacteria have only a limited effect on stone surfaces.

## 8.2 Algae, Fungi, Lichens, and Mosses

We have evidence that the low forms of life, algae, fungi, lichens, and mosses, extract ions by the production of carbonic, nitric, sulfuric, and other weaker acids. Silvermann and Muñoz (1970) studied the activity of the common fungus *Penicillium simplicissimus* in vitro. Si, Al, Fe, and Mg are released by nitric or sulfuric acid. The degree of solubilization or leaching of

**Fig. 8.2.** Concentration of the orange fungus *Hematococcus pluvialis* populates cracks of white Vermont marble. The organisms have penetrated 1.5 mm into the marble. High precipitation and relative humidity favor such growth. Amphitheater, Arlington National Cemetery, Washington DC

Si ranged from 0.3 to 31%, of Al from 0.7 to 11.8%, but of Fe from 25 to 60%; the pH of the initial solution dropped from 6.8 to 3.5 in 7 days. Basic igneous rocks with a low silica content are generally more easily attacked than are acid granitic rocks, which remain quite resistant to fungal attack. Iron and magnesium record the greatest loss in all rocks, while Si and Al were removed only from basic igneous rocks. Regardless of dissolution, citric acid can also chelate Ca from carbonate rocks. Soot and a high amount of urban sulfate on a stone surface prevent lichens and mosses from settling. Occasionally, marble is populated by a fungus with a bright orange pigment; the fungus, *Hematococcus pluvialis*, was observed by this writer mostly along cracks of white Vermont marble (Fig. 8.2). The same feature was observed previously on white Carrara marble by Realini et al. (1985), although they did not identify the organisms. Such fungi readily populate stone surfaces without association with sulfur, iron, or nitrogen. All organisms secrete organic acids and organic decay residues, which accelerate stone corrosion.

Boring by microorganisms in a marine environment has been sketched by Golubič et al. (1975; see Fig. 8.4). They describe three types of lifestyle: deep borings, mostly by algae (endolithic); shallow trenching into the rock substance (chasmolithic); and surficial growth (epilithic), leaving an etched surface. Deeper borings may become the home for more advanced animals which carry out further boring.

Some marine algae bore into the rock substance by the action of end cells of endolithic filaments. The algae prefer cleavage planes and grain boundaries (Golubič 1969).

The symbiotic nature of lichens, algae, and fungi is generally characterized by slow growth and the ability to attach to bare rock surfaces without the need of supporting soil (see Fig. 8.8). They are the pace-makers in the formation of humus which, in turn, supports higher plants. Discoloring of translucent calcites can be observed. Silicolous lichens in silicate rocks live either within the rock or on the surfaces, either encrusting the stone surface, often in patches, or as foliose lichens with leaf-like lobes. Wainwright (1986) names four crustose and eight foliose lichens observed growing on a vertical 100-ft-high granite surface of the Walt Whitman memorial near Cloyd, Ontario, Canada; he gives the names and characteristic colors. Stone surfaces are affected by lichens by:

1. Mechanical water retention. Spongy lichens can retain water for a long time.
2. The secretion of organic acids. Humic and other organic acids are produced from the organic remains which readily attack by dissolution and chelation, leading to accelerated decay. Acid-reacting $H^+$ continues its attack on the unweathered mineral substance beneath. Schatz et al. (1954) have observed that chemical degradation of rock and stone appears to be mediated by chelation. Acid strength is not an index to the chelating ability. Lichens appear to be the most prolific producers of chelators, some of them replacing the function of soil. Natural residual soils contain many chelating agents which accelerate the breakup of rock and stone and provide nutrients for plants. Humid tropical areas are especially endangered by the huge population of microbes. The Buddhist monument of Borobudur on Java, Indonesia, is a well-studied example of the multitude and quantity of microbial life on carvings in porous volcanic rock (Table 8.1).

Quantitative laboratory experiments with fungi and lichens on minerals and rocks were performed by Henderson and Duff (1963). The fungi *Aspergillum niger*, *Spicaria* sp., and *Penicillium* sp. produced considerable quanti-

**Table 8.1.** Microbial analysis for the Temple of Borobudur

| Microfauna | Soil sample (topsoil) | Stone sample | |
| --- | --- | --- | --- |
| | | Near ground | 1 m above ground |
| Bacteria, total | 130 000 | 16 000 000 | 250 000 |
| Fungi | 3 000 | 230 000 | 13 000 |
| Actinomycetes | 100 000 | 3 000 | 30 000 |
| Algae | 6 000 | 90 000 | 6 000 |
| Nitrogen bacteria | 1 475 | 12 500 | 12 100 |
| Sulfur bacteria | 183 | None | None |

After *Voûte (1969)*; all values are number of life-forms per gram of rock.

**Table 8.2.** Microbial population connected with surface scales

| Alveolization (honeycombs) | | Spalls |
| --- | --- | --- |
| Microflora, total | $2.5 \times 10^7$ | $1.1 \times 10^4$ |
| Sulfate (%) | 4.77 | 13.4 (mostly by *Thiobacillus*) |

ties of citric and oxalic acid from a 4% glucose solution. Such acids are believed to be most active in stone decay. Al, Mg, and Si were extracted from a few rock-forming minerals exposed to fungal culture solutions. The most active fungus appears to be *Aspergillum niger*, a common fungus in some black lichens which removed 14% of $SiO_2$. The weathering rate of silicate rocks can therefore be accelerated under a cover of this common fungus.

Honeycombs can develop a sizeable amount of a microbial population on and within surface scales (Jaton 1972) (Table 8.2).

## 8.2.1 Desert, Land

The microbial population is generally high in moderate semihumid and humid climates, but much higher in humid tropical regions. The greatest microbial activity in stone appears near the ground, above the soil level. Microbial chemical activity is summarized in Table 8.3. Microorganisms are also found in hot and cold deserts. Their presence, identified by metabolism as Biological Oxygen Demand (BOD), has been detected on the Egyptian temples in Karnak, Egypt, by Curri and Paleni (1975). But Danin et al. (1982) have determined a terrestrial cover of blue–green algae (Cyanophytes) populating the stone surface with a maximum penetration to 0.3 mm at 500 mm annual precipitation; at only 100 mm rainfall the algal penetration can be 30 mm. Desert conditions often prevail on urban masonry walls in a semihumid climate. Shachak et al. (1987) describe rock weathering in the

**Table 8.3.** Summary of microbial life and its geological activity (After Silverman and Ehrlich 1964)

| Element | Microorganism | Geological activity |
|---|---|---|
| Ca | *Arthrobacter* | Dissociate calcite, dolomite |
| Mg | *Thiobacteria* | Attacks Mg only |
| | *Desulfofibrio* | Attacks Mg only |
| Fe oxidation | *Thiobacter ferrooxidans* | Oxidizes ferrous ions |
| | *Ferrobacillus* | Oxidizes ferrous ions |
| Fe reduction | *Desulfofibrio* | Reduces ferric to ferrous ions |
| N | *Azotobacter* and others | Calcite to Ca-nitrate, gypsum |
| P | *Penicillium* and others | Dissociate Ca-phosphate |
| S oxidation | *Thiobacteria* fungi, filamentous | Sulfur to sulfates from the atmosphere |
| S reduction | *Desulfofibrio* yeast | Reduces sulfates to $H_2S$ at or near ground level |
| Si | Bacteria, fungi | Dissolution of silicates |
| | Actinomycetes | |
| | *Aspergillum* | |
| | *Penicillum* | |

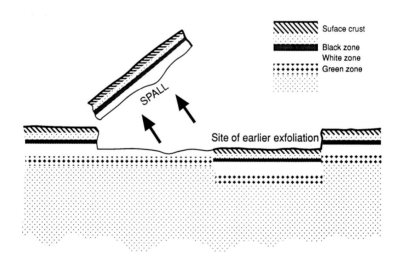

**Fig. 8.3.** Exfoliation and spalling by algal growth (blue–green algae) on sandstone. Black zone represents decayed algal growth, and the green zone active algal growth. (Adapted from Friedmann 1982)

limestones of the Negev desert, Israel, caused by endolithic algae and foraging snails which cause considerable surface attack. The limestone surface can resemble a jigsaw or bread-crust pattern. Exfoliation in sandstones on masonry walls may also be affected by such action. Friedmann's (1982)

| on stone surface | near surface | within stone |
| (EPILITHIC) | (CHASMOLITHIC) | ( ENDOLITHIC) |

**Fig. 8.4.** Sketches of surface algal activity (epilithic), shallow boring (chasmolithic), and deep boring (endolithic) activity in stone surfaces. (Adapted from Viles 1987)

observations appear important for microbial rock scaling (Fig. 8.3). Urban masonry walls tend to reflect conditions in deserts.

### 8.2.2 Marine Environment

Marine microphytes have attacked limestone in a shallow marine environment (Lukas 1979) to a depth of 370 m, with the greatest efficiency at the intertidal zone, where there is a rate of decay of 0.3 to $36 \mu m$/day. Viles (1987) describes blue-green algal attack in marine intertidal zones of a tropical coral reef; this was supported by the exposure of test tablets in this zone. The author claims great complexity of the biogenic weathering processes. Submersed monuments and sea walls are frequently attacked by endolithic filamentous microboring fungi and algae living inside (Fig. 8.4).

## 8.3 Biochemical Action of Higher Plants

The dissociation of calcite and silicates by processes other than acidity (carbonic and organic acids) is known by soil scientists. Keller (1957) shows that calcite dissociates in an aqueous solution of Na-EDTA by chelation, during which reaction Ca is taken up, despite pH values of 10 and above, when no additional calcite attack can be expected. Chelation is the uptake of metal ions into a void space of an organic ring structure in which, e.g., Ca is held tightly. Keller (1957) sketches the interaction of plant roots with a silicate mineral grain which emits $H^+$ ions from the active root tip (Fig. 8.5). The $H^+$ cation produced by the rhizome of lichens and by the root tips of more advanced plants readily exchanges with negatively charged nutrient metal cations within the minerals and soils. Climbing vines on buildings, like Boston ivy, Virginia creeper, trumpet vine, and others, leave their marks from rootlets and attachment cups on the masonry and mortar. Such remains

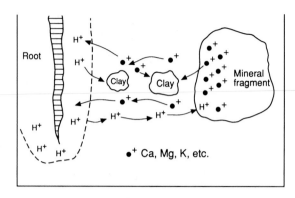

**Fig. 8.5.** Biochemical interaction of a plant root with a silicate mineral grain. (Keller 1957)

tend to produce corrosive organic acids, and also prevent drying of the masonry beneath.

## 8.4 Physical Plant Action

Ferns, grasses, and small trees tend to find a foothold on buildings; they are supplied with nutrients from simple plants, like bacteria and algae, on and between the stone blocks. The growing plant can exert high growth pressures, destroying and wedging along crevices and mortar joints on older buildings, especially in areas of ample moisture supply. Plant-root wedging along rock cracks is well known. Vegetation interfering with historic structures was discussed by Goeldner (1984). The pressure of a growing plant root or a tree trunk is surprisingly high. Gill and Bolt (1955) distinguish between axial and radial pressures of plant roots for corn, beans, and peas. Though such crop plants are not associated with building stones, they give an indication of potential axial and radial pressures.

Axial pressures on root tips are measured along the long axis of the root; maximum pressures of about 19 atm (278 psi) were recorded along bean root tips.

Radial pressures of roots are perpendicular to the long axis, and are one-third to one-fourth of the axial pressure. The pressures become effective as soon as growth is inhibited. Figure 8.6 shows a quartzitic millstone fractured by the radial pressure of a growing tree trunk. The effective root and trunk surface is much larger for radial pressure than for axial pressure: the wedging force is therefore high, causing cracks and joints to widen and allow easier access for moisture. The root pressure equals the plant osmotic pressure in growing cotton, peas, and peanuts. The approximate root pressure is about 10% lower than the measured osmotic pressure of a

**Fig. 8.6.** Abandoned millstone broken by a growing tree trunk. Spring Mill State Park, Indiana

plant (Taylor and Ratcliff 1969). A continuous average radial pressure of 15 atm (220 psi) should be expected by advancing plant roots and root-lets. Roots tend to grow along rough stone surfaces, where they interact chemically with the minerals to extract nutrients through surface attack while they seek a weak joint or crack.

## 8.5 Boring by Advanced Animals

Intertidal marine zones, the shoreline between the high tide and the low tide, are endangered by rock borers. This is the zone above manmade sea walls of stone or concrete, or partly submerged historical monuments. The temple of Jupiter Seraphis on the Mediterranean shore near Naples, Italy, is marked by two distinct rows of densely spaced boring mussels, located today about 30 ft above the present sea level (Fig. 8.7). There is evidence that the temple was submerged twice into the intertidal zone during the last 1800 years (Dvorak and Mastrolorenzo 1991). The systematic description of distinctive excavation characteristics for boring fungi, algae, sponges, poly-chete worms, snails, bivalve clams, and sea urchins is given by Warme (1975). Most borers dig for protection (Fig. 8.8). Deep and closely spaced

**Fig. 8.7.** Holes of boring clams high on the columns of the ancient Seraphis temple near Naples, Italy. The temple was built in 105 B.C.; the land sank and rose several times through the Middle Ages. The stable ocean level is out of sight today. Reproduced from an engraving in the late 1700s. (Dvorak and Mastrolorenzo 1991)

borings roughen the stone surface and can weaken the stone structure, often to the limit of bearing strength.

Boring is achieved either by mechanical abrasion or by chemical etching (acid secretion). Ansell (1969) describes the mechanical boring mechanism as follows: The animal anchors the end of the strong foot, near the exit of the hole, against the wall by expanding its foot, while the sharp-edged shell abrades the stone by twisting back and forth around its axis. Various species of angel wings (*Pholas*) and false angle wings (*Petricola*) are mechanical rock borers in carbonate rock, concrete, and soft shale. Angel wings dig at a rate of 12 mm/year, reaching a total depth of about 150 mm. Shore erosion on the US west coast can be substantial (Evans 1986). Warme (1969)

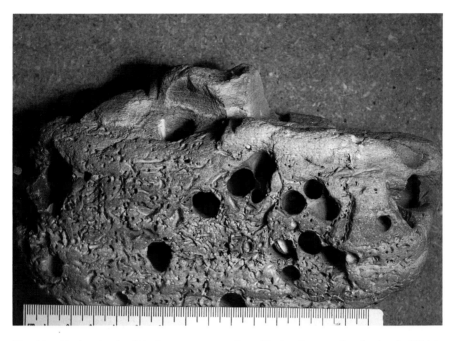

**Fig. 8.8.** Algal and animal borings on stone surface. Depicted are surface boring (epilithic), shallow borings by worms (chasmolithic), and deep borings by bivalve clams. Block from Malibu Beach, California

observed a higher population density of pholad boring clams in soft rock than in hard material. Sea urchins of the genera *Echinus* and *Eucidaris* are able to bore by sinking their sharp teeth, at the center of the bottom of their body, into the rock underneath. The rate of their boring can be as much as 1 cm/year in limestone, and faster in softer rock (Fig. 8.8).

Chemical borers are restricted to carbonate rocks. Boring clams should be considered the initial borers, followed by other boring organisms which dig an intricate network of channelways of various sizes and leading to the final disintegration of the rock substance. The clam *Lithophaga* digs straight smooth channels to a depth of 100 mm. Boring mussels, *Mytilus*, are the most common of the advanced boring animals found in limestone (Fig. 8.7). Following the action of clams, polychete worms continue to dig a network of small round or oval channels. Densely spaced, the sponge *Cliona* completes the work of rock fragmentation to a depth of 25 mm, whereby the friable debris crumbles as sand to the seafloor (Neumann 1966). The primitive sponge *Cliona* bores into the rock substance by advancing plasma as "amoebocytes", whereby the main part of the cell with the cell nucleus remains in place after an outline has been etched into the rock (Cobb 1969). The observed marine bioerosion on tropical Bermuda is as given in Table 8.4:

**Table 8.4.** Marine bioerosion. (From Neumann 1966)

| | |
|---|---|
| Below the tidal zone | 1.4 cm by the sponge *Cliona* |
| | 1.3 cm by the boring clam *Lithophaga* |
| Intertidal zone | Little erosion by browsers scraping |
| | for algal growth |
| Above tides | No bioerosion found |

Bioerosion can reach catastrophic rates of up to 25 times the normal measured rates, aided by marine solution and mechanical wave action. Warme (1975) has summarized bioerosion.

## 8.6 Birds on Buildings

Urban buildings are often inhabited by a variety of birds seeking a place for rest, refuge, or breeding beneath projections, pinnacles, and other protected spots. Pigeons and the common English sparrow are the most common birds best adjusted to man's activities. House martins (swallows) are frequent inhabitants of suburban areas and underneath bridges. Large quantities of excrement release phosphoric and nitric acids, which etch stone and chemically react with carbonates to form calcium phosphates and nitrates. It is difficult to evaluate and estimate the participation of birds.

Protection against this annoyance can be relatively simple, ranging from shooting or poisoning birds, using recorded noises, to generally acceptable protective nets of almost invisible nylon. Groups against animal cruelty are limiting factors in bird control.

## 8.7 Control of Biological Growth

The removal of stone surfaces infected with microbial life, bacteria, algae, and lichens, should first be attempted using the easiest methods available, such as chlorine (Clorox-type) bleaches of 1:1 dilution with water, possibly with application by poultice for a prolonged reaction. Bleaching peroxide ($H_2O_2$) and dilute ammonia are also recommended as they tend to oxidize the organic living cell (Grimmer 1988). The removal of encrusting lichens from a granite surface was successfully achieved by Wainright (1986), who used three herbicides: a 5% (w/v) solution of sodium hypochlorite, denatured ethanol (ethyl alcohol), and a 0.1% (w/v) solution of ortho-phenylphenol in denaturated ethanol. Lysol was used for smaller surfaces. Richardson (1973) describes the cleaning of stone surfaces, especially to remove algae and mosses on tombstones and roofs in humid England. He warns of the use of cleaning agents that can combine with metal ions to form new efflorescing compounds, discoloring light-colored stone, especially white marble. Thaltox Q, and Gloquat C (ammonium compounds) have been used successfully. A large surface may have to be treated with chemicals in a highly diluted form, and treatment repeated with the changing growing seasons, spring and fall in the Northern Hemisphere. Weedkillers were modified to kill bacterial and algal tropical flora at the Buddhist temple at Borobudur, Java. Valuable bas reliefs extending over 4.8 km in volcanic rock were successfully treated with Hyvar-X, a common weedkiller produced by the DuPont company; this has been successful in eliminating algae, lichens, and mosses. Sufficient residual chemical was left on and within the stone surface to prevent future growth (Schmidlin 1979). Lethal irradiation with low-wavelength UV, peaking at 253 nm, is recommended by Van der Molen et al. (1980). UV radiation has been applied in the canning industry for some time to successfully kill all microbes for complete sterilization before the cans are sealed. Caution should be taken because this radiation is carcinogenic to the human skin. Such radiation was arranged with commercially available lamps mounted on fluorescent fittings. The radiation does not penetrate the stone substance. Much experimentation is still in progress to find an efficient, safe, and lasting method of ridding stone of microbial life.

# References

Ansell AD (1969) A comparative study of bivalves which bore mainly by mechanical means. Am Zool 9:857–868

Berthelin J (1983) Microbial weathering processes. In: Krumbein WE (ed) Microbial geochemistry. Blackwel Oxford, pp 233–262

Cobb WR (1969) Penetration of $CaCO_3$ substrates by the boring sponge Cliona. Am Zool 9:783–790

Curri SB, Paleni A (1975) Some aspects of the growth of chemolithotrophic microorganisms on the Karnak Temples. The conservation of stone, vol I. In: Rossi-Manarasi R (ed) Proc Int Symp Bologna, 19–21 June, pp 267–279

Danin A, Gerson R, Marton K, Garty J (1982) Patterns of limestone and dolomite weathering by lichens and blue–green algae and their paleoclimatic significance. Paleogeogr Paleoclimatol Paleoecol 37:221–233

Dvorak JJ, Mastrolorenzo G (1991) The mechanism of recent vertical crustal movements in Campi Flegrei caldera, southern Italy. Geolog Soc Am Spec Pap 263:47

Evans JW (1986) The role of Penitella penita (Family Pholadidae) as eroders along the Pacific coast of North America. Ecology 49(1):156–159

Friedmann I (1982) Endolithic microorganisms in the Antarctic cold desert. Science 215:1045–1052

Gill WR, Bolt GH (1955) Pfeffer's study of the root growth pressures exerted by plants. Agron J 47:166–168

Goeldner PK (1984) Plant life at historic properties. Assoc Preserv Technol Bull XVI(3/4):67–69

Golubič S (1969) A comparative study of bivalves which bore mainly by mechanical means. Am Zool 9:857–868

Golubič S, Perkins RD, Lukas KJ (1975) Boring microorganisms and microborings in carbonate substrates. In: Frey RW (ed) The study of trace fossils. Springer, Berlin Heidelberg New York, pp 2329–2359

Grimmer A (1988) Keeping it clean. US Dep Inter Natl Park Serv 34 pp

Henderson MEK, Duff RB (1963) The release of metallic and silicate ions from minerals, rocks and soils by fungal activity. J Soil Sci 14(2):236–246

Iverson WP (1974) Microbial corrosion of iron. In: Microbial iron metabolism. Academic Press, London, pp 475–513

Jaton MC (1972) Aspects microbiologiques des alterations des pierres de monuments. Ier Colloq Int sur la Deter de Pierres, 11–16 Sept 1972, LaRochelle, pp 149–154

Kauffmann J (1960) Corrosion et protection des pierres calcaires des monuments. Corrosion et Anticorrosion 8(3):87–95

Keller WD (1957) The principles of chemical weathering. Lucas, Columbia, Missouri, 111 pp

Lukas KJ (1979) The effects of marine microphytes on carbonate substrata. Scanning Electron Microsc 1979(II):447–456

Mamillan M (1968) L'alteration et la preservation de pierres dans les monuments historiques. Etude de l'alteration des pierres, vol I. Colloq tenus a Bruxelles le Ferr 1966–1967, pp 65–98, Cons Int des Monuments et des Sites, ICOMOS

Millot G, Cogne J et al (1967) La maladie des gres de la Cathedral de Strasbourg. Bull Serv Carte Geol Elsace-Lorraine 20(3):131–157

Neumann AC (1966) Observation of coastal erosion in Bermuda and measurements of the boring rate of the sponge *Cliona lampa*. Limnol Oceanogr 1(1):92–108

Pochon J, Jaton D (1967) The role of microbiological agencies in the deterioration of stone. Chem Indust 47:1587–1589

Realini M, Sorlini C, Bassi M (1985) The Certosa of Pavia: A case of biodeterioration. In: Felix G (ed) Vth Int Congr on Deterioration and conservation of stone, Lausanne, 26–27 Sept

Richardson BA (1973) Control of biological growth. Stone Indust 8(2):2–6

Schatz A, Cheronis ND, Schatz V, Trelawny GS (1954) Chelation (sequestration) as a biological weathering factor in pedogenesis. Proc Penn Acad Sci XXVIII:44–51

Schmidlin W (1979) Rescuing an architectural wonder. DuPont Magazine Nov/Dec 1979:1–10

Shachak M, Jones CG, Granot Y (1987) Herbivory in rocks and the weathering in a desert. Science 236:1098–1099

Silverman MP, Ehrlich HC (1964) Microbial formation and degradation of minerals. Adv Microbiol 6:153–206

Silverman MP, Muñoz EF (1970) Fungal attack on rock: Solubilization and altered infrared spectra. Science 169:985–987

Stolzy LH, Barley KP (1968) Mechanical resistance encountered by root growth habits of plants. Soil Sci 105:297–301

Taylor HM, Ratcliff LF (1969) Root growth pressure of cotton, peas, and peanuts. Agron J 61:398–402

Van der Molen J, Garty J, Aardema B, Krumbein WE (1980) Growth control of algae and Cyanobacteria on historical monuments by a mobile UV unit (MUVU). Stud Conserv 25:71–77

Viles HA (1987) Blue-green algae and terrestrial limestone weathering on Aldabra Atoll. Earth Surface Processes and Landforms 12:319–330

Voûte C (1969) Indonesia; geological and hydrological problems involved in the preservation of the monument of Borobudur. UNESCO Rep Ser No 1241/BMS RD/CLT, Paris, May 1969, 37 pp

Wainwright NM (1986) Lichen removal from an engraved memorial to Walt Whitman. Assoc Preserv Technol Bull XVIII(4):46–51

Warme JE (1969) Marine borers in calcareous rock of the Pacific coast. Am Zool 9:783–790

Warme JE (1975) Borings as trace fossils and the process of marine bioerosion. In: Frey RW (ed) The study of trace fossils. Springer Berlin Heidelberg New York, pp 181–227

Webley DM et al. (1963) The microbiology of rocks and weathered stones. J Soil Sci 14(1):102–112

# 9 Iron in Minerals and the Formation of Rust in Stone

The mean iron content of the earth's crust is 5%. Iron is locked in ferromagnesian silicates in rocks at the earth's surface mostly as green or black ferrous-ferric iron. The black ferrous-ferric form is magnetite, the red ferric oxide, hematite, and the yellow-brass ferrous sulfides are commonly cubic pyrite and orthorhombic spearhead-shaped marcasite. Iron also appears as white to dark brown ferrous carbonate (siderite) and green iron silicate, glauconite, which adds a greenish color to sedimentary rocks (Sect. 4.4).

The red, maroon to deep red ferric oxide (the most powerful pigment of sedimentary rocks) can accumulate locally to form iron ore deposits of commercial importance. Humid to semihumid climates readily oxidize metallic iron and iron minerals to limonite (ferric hydroxide), common rust. Limonite is a mixture of finely crystalline goethite (the alpha-FeOOH) and the amorphous form of the same composition and color. Metallic iron also changes to rust with time, mostly amorphous ferric hydroxide. Crystallization by aging of the noncrystalline ferric hydroxide leads to the submicroscopic goethite and the deep orange lepidocrocite (Fig. 4.9).

The stability of the iron minerals and minerals containing iron in their crystal lattices are of great importance to stone color. Figure 9.1 gives a general overview of the stability of iron compounds as a function of pH, acidity, and the Eh (oxidation-reduction) of the environment. Pyrite ($FeS_2$), the most common and important constituent, has caused extensive damage.

## 9.1 Pyrite, Marcasite

Both sulfides of iron occur in almost all types of rock, and sometimes in considerable quantities. Pyrite can occur as single small crystals, as coatings, or as vein fillings in all rocks. The gold to brass color of the unweathered mineral makes it a rather desirable-looking constituent. Pyrite is deep black as a powder, the color of amorphous ferrous sulfide. Both minerals weather equally fast to rust in only a few years when exposed to the atmosphere, oxidizing and simultaneously releasing sulfur as sulfate. This ion in the presence of water generally converts to sulfurous acid or to the much more corrosive sulfuric acid. The corrosive action of the acids can be readily observed in marbles where in situ veins or patches of pyrite are exposed to

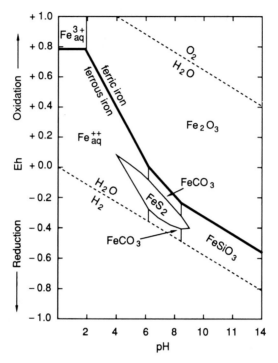

**Fig. 9.1.** Stability range of the common iron minerals as a function of pH and oxidation – reduction, the Eh of the environment. Compare the stability range of pyrite with the observations of pyrite weathering. (Garrels and Christ 1965)

the weather. Figure 9.2 shows the considerable surface reduction in Vermont marble on both sides of a quartz-pyrite vein. The developing rust may be reduced again by the remaining unweathered pyrite. The stability environment is reducing and shows a mid-range pH from 4 to 8. Buffering by carbonate environments does not easily shift pyrite from its stability range. Microorganisms, like *Thiobacillus thiooxidans* and others, do not effectively oxidize pyrite above pH 4. Arkesteyn (1980) found in experiments that more than twice the amount of sulfur became available during 140 days of exposure in noncalcareous compared with calcareous soils. In urban areas acid rain on carbonate rocks is quickly buffered from less than pH 4 to neutral values (Figs. 5.1 & 9.1). The disintegration process of pyrite, simplified after Weber (1966), is as follows:

$$2FeS_2 + 7O_2 + 16H_2O = 2FeSO_4 \cdot 7H_2O + 2H_2SO_4$$

$$FeSO_4 \cdot 7H_2O = FeSO_4 + 7H_2O$$

$$12SO_4 + 3O_2 + 6H_2O = 4Fe_2(SO_4)_3 + 4Fe(OH)_3$$

$$Fe_2(SO_4)_3 + 6H_2O = 2Fe(OH)_3 + 3H_2SO_4.$$

**Fig. 9.2.** Surface reduction of crystalline marble between two narrow pyrite veins. The reduction is greater between the veins than on the marble surface. The iron was reduced by the pyrite to a green tarnish of the ferrous form, following 40 years of exposure to 900-mm annual precipitation in a South Bend cemetery

The process of oxidation–hydration can be more complex and is outlined in four consecutive stages.

Stage 1: $FeS_2$ (solid) plus $O_2$ goes to $Fe^{2+}$, plus free sulfur. The solid ferrous sulfide takes up oxygen in contact with rainwater and the atmosphere, leading to both ionized ferrous and ferric ions plus the sulfate ion. The process continues as oxidation of the ferrous iron.

Stage 2: $Fe^{2+}$ plus $O_2$ (aqueous) oxidizes to $Fe^{3+}$. Oxidation of the ferrous iron ($Fe^{2+}$) takes place by the oxygen present in rainwater after migration has occurred for a short distance.

Stage 3: $Fe^{3+}$ plus $H_2O$ precipitates as $FeOOH$ (solid rust). $Fe^{3+}$ precipitates first as amorphous ferric hydroxide, which crystallizes slowly to

**Fig. 9.3.** Strong discoloring of gray Indiana limestone on a suburban residence. Veins of pyrite in the limestone have weathered to brown limonite (rust), weeping downward to the ground, despite the strong buffering action of the limestone. Indiana limestone ordinarily does not show such extensive pyrite veins. It appears that a contractor bought up a lot of stone blocks which were discarded on a waste pile

the microcrystalline ochre-brown mineral goethite. The solubility of rust, ferric hydroxide, is only about 0.0005 ppm at pH 5, and even less at pH 7. The weathering of pyrite releases sulfuric acid, lowering the pH of the attacking rainwater to near pH 3; at this acidity the solubility of rust increases to 5 ppm, and increases further to 550 ppm at pH 2. The observed brown halos of rust at some distance from the source are contrary to the very limited solubility of rust near a neutral pH. The presence of iron and sulfur bacteria is believed to greatly accelerate oxidation and facilitate the development of rust from pyrite by a factor of 1 million (Singer and Stumm 1970). The discoloring of Indiana limestone by rust from pyrite veins is surprising and should be a warning (Figs. 9.3 & 9.4).

Stage 4: $Fe^{3+}$ plus $FeS_2$ (solid) reduces back to $Fe^{2+}$ plus $SO_4^{-2}$, forming a green tarnish on a strongly reducing pyrite surface, provided that the pyrite masses are large enough for the precipitated rust to remain in contact with the fresh parent pyrite (Fig. 9.2). None of the veins or patches which were filled with pyrite appear to be cracked by up to 50% expansion with rust formation (Table 9.1). The temporary acid environment of the weathering by-product, the sulfuric and related acids, has retained the ferrous form in solution long enough to travel the distance as ferrous iron. Sunlight

**Fig. 9.4.** Dark brown rusty patches weeping downward from nodules of pyrite enclosures in cream local limestone. Country church in Weeping Water, Nebraska. After original color photo by Prof. William J. Wayne, Dept. of Geology, University of Nebraska, Lincoln, Nebraska

**Table 9.1.** Volume increase in iron and iron minerals with conversion to rust

| Original material | Density (g/cm³) | Expansion to goethite (%) | Expansion to rust (%) |
|---|---|---|---|
| Pure iron | 7.88 | 186 (D = 4.37) | 200–250 |
| Structural steel | 7.50 | 164 | 150–250 |
| Goethite | 4.37 | – | 20–70 |
| Pyrite ($FeS_2$) | 5.02 | −15 | 20–50 |
| Marcasite ($FeS_2$) | 4.89 | −17 | 20–50 |

enhances the reducing action and the solubility of iron by photoreduction by a factor of 4 compared with night-time activity (McKnight et al. 1988). Bacterial action may also influence the solubility, possibly with the help of chelation.

Protection from decay is discussed by Oddy (1977) for museum specimens which tend to decay rapidly by oxidizing iron sulfides. Treatment with ammonium vapor neutralizes a part of the developing acid, sealing the stone objects. The stone surface is protected from the access of moisture and oxygen. The use of most bactericides can also slow the oxidation process considerably.

## 9.2 Ferrous-Ferric Iron in Minerals

### 9.2.1 Biotite

The black mica loses its ferrous-ferric iron from the silicate lattice at the beginning of the weathering process and soon precipitates as rust halos around the mica flakes. The release of iron from biotite in distilled water is almost faster than leaching from the other ferromagnesian silicates. Laboratory data by Keller et al. (1963) correspond with observations made for buildings and monuments.

### 9.2.2 Hornblende, Augite

These minerals contain a variable amount (7–15%) of black to green iron in the ferrous form. The release of iron is very slow. It may take several generations for discoloring to occur.

### 9.2.3 Siderite

Ferrous carbonate occurs as occasional grains that are light gray to brown. Upon dissolution in water the iron oxidizes readily to the ferric form, similar to pyrite (Weber 1966). The yellowing of many crystalline white marbles may originate from the oxidation of scattered small grains of siderite. Such oxidation may have occurred in the Pentelian marble on the Parthenon in Athens. Localized siderite nodules enclosed in exposed concrete aggregate have developed halos of ferric hydroxide that point away, by 3 to 4 in., from the wind-driven direction. They are also often seen on concrete with exposed gravel in the northeastern part of the USA.

### 9.2.4 Hematite, Magnetite

Hematite, the alpha-$Fe_2O_3$, and the magnetic $Fe_3O_4$ are stable minerals. Hematite in crystalline form is much more stable than the amorphous or microcrystalline form. Hematite and magnetite occur in granites; more frequently in dark igneous rocks. Small residual grains of hematite or magnetite are occasionally found in sandstones. Reddish dusts from operating iron smelters tend to settle on nearby buildings which are protected from washing rain. The thin crust of red dust may be intermixed with soot, as is also industrial dust containing some metallic iron and iron alloys. Many industrial dusts are rich in hematite and magnetite; some of the iron may migrate across the stone surface.

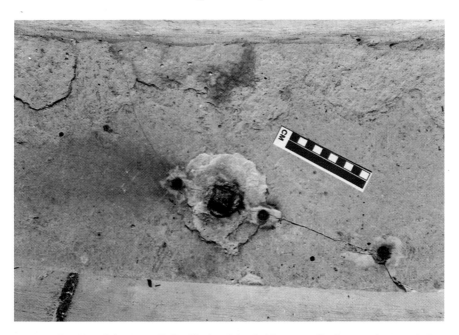

**Fig. 9.5.** Cracking of stone step (Joliet Silurian dolomite) by sawn-off railpost and support bolts. Ferric iron as rust has stained the zone around the iron pegs. Main entrance to Sacred Heart Church, Notre Dame, Indiana

## 9.3 Damage to Stone by Rust

Steel anchors and bolts inserted into stone or concrete have caused damage by rust burst wherever there is access to the atmosphere (Fig. 9.5). Similar rust burst can be observed by the oxidation of pyrite, or indeed wherever rust can develop in the presence of oxygen and moisture. Theoretical calculations of the volume expansion of inserted rusting bolts, pins, and anchors from metallic iron and iron minerals is presented in Table 9.1.

The exact calculation of the volume expansion from metallic iron to rust is very difficult because most freshly formed rust is a heterogeneous mixture of amorphous and crystalline $FeOOH$ with different degrees of hydration. The theoretical calculations of the volume expansion of the important weatherable iron minerals are approximations and guidelines. Gottlieb (1988) recommends how to find, remove, and replace hidden clamps, cramps, and pins with a magnetometer, if they are damaged by rust, and how to expose them by core drilling for repair, installation of weep holes, creating space for expansion by rust formation, and treatment of still intact iron enclosures with rust-proofing paint.

### 9.3.1 Matching of Replaced Stone

White marble and other light-colored stone in yellowed patina-covered masonry may display an undesirable patchwork of new stone between old stone.

An aqueous solution of ferrous chloride, almost clear when painted on the white stone surface, turns ochre after a few hours. It adheres to the stone surface like nature's patina. The process can be repeated till the proper color match is reached. Ferrous chloride is nontoxic and very soluble in pure water.

# References

Arkesteyn GJMW (1980) Pyrite oxidation in acid sulfate soils: The role of microogranisms. Plant Soil 54:119–1334

Garrels RM, Christ CL (1965) Solutions, minerals and equilibria. Harper and Row, New York, 450 pp

Gottlieb SE (1988) Preventing limestone spalls, McKim, Mead and White buildings at Columbia University. Assoc Preserv Technol Bull XX(3):33–27

Keller WD, Balgoard WD, Reesman AL (1963) Dissolved products of pulverized minerals, Part I. J Sediment Petrol 33(1):191–204

McKnight DM, Kimball BA, Bencala KE (1988) Iron photoreduction and oxidation in an acidic mountain stream. Science 240:637–640

Nicholson RV, Gillham RV, Reardon EJ (1990) Pyrite oxidation in carbonate-buffered solution. 2. Rate control by oxide coatings. Geochim Cosmochim Acta 54:395–402

Oddy WA (1977) The conservation of pyritic stone antiquities. Stud Conserv 22:68–72

Singer PC, Stumm W (1970) Acidic mine drainage: the rate-determining step. Science 167: 1121–1123

Weber J (1966) Die Verfärbungen an natürlichen Bausteinen infolge Verwitterung von Eisensulfidmineralien. Schweiz Bauztg 84(5.3):1–7

# 10 Fire Resistance of Minerals and Rocks

Urban fires have severely damaged stone in the past. Most affected are granites, quartz sandstones, limestones, dolostones, and marbles. Major conflagrations in many European cities during World War II left numerous ruins for the study of damage to stone. The behavior of minerals and rocks at high temperatures principally resembles solar heating–cooling cycles of rock and stone; these can lead to cracking and warping of thin stone panels. These are relatively slow processes compared with the spontaneous heating in major fires (Sect. 2.11). Uneven volume (linear) expansion of minerals may cause the disruption of stone and concrete during a fire. All minerals expand when heated.

Damage to stone by fires was studied both in the field and on urban buildings. The curve for heating in a closed room of a "standard fire" within a closed building was measured by the Underwriter's Laboratories (UL). The UL curve resembles the German DIN (Deutsche Industrie Norm) curve (Fig. 10.1). Heating of rock-forming minerals results in uneven expansion of the minerals (Fig. 10.2). Three important temperatures are marked in a "standard fire" in an enclosed room (Underwriter's Laboratory 1964): 538 °C after 5 min, 704 °C after 10 min. The temperature of the phase change of quartz at 573 °C from low to high quartz with nearly 4.5% volume expansion, is therefore the most disruptive of the common rock-forming minerals in a fire. The extensive damage by major conflagrations during World War II supplied ample material for the study of damage to stone. The German DIN fire stability curve for building materials appears to be a workable basis for anywhere in the world (Fig. 10.1; Kieslinger 1954).

## 10.1 Expansion of Minerals at High Temperatures

The volume change at elevated temperatures is given in Fig. 10.2 and the linear change is given in Fig. 10.3 for the important rock-forming minerals, quartz, calcite, orthoclase, plagioclase, and hornblende, up to 800 °C.

*Quartz* expands about four times more than plagioclase and feldspar, and twice as much as hornblende (Fig. 10.2). Even more damage may occur by the often great contrast of linear expansion, the expansion along the long axis versus the expansion of the short axis. This contrast may cause excessive

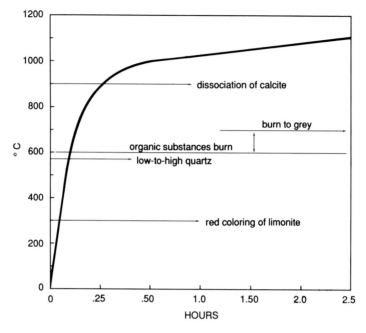

**Fig. 10.1.** Standard heat curve in a closed room. From UL #263 (1959) and DIN DVM 4102. (Kieslinger 1954)

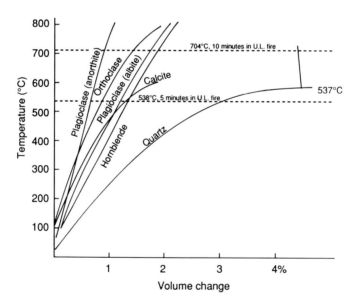

**Fig. 10.2.** Volume expansion of some important rock-forming minerals. The UL Standard plots the temperature for 5 min heating at 538 °C and 10 min at 705 °C. Data plotted from Skinner (1966)

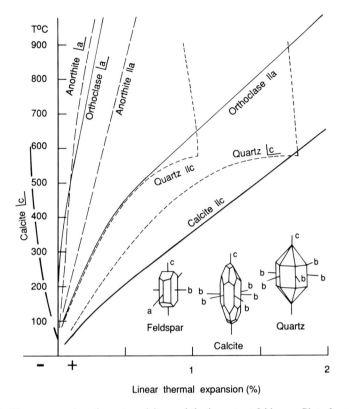

**Fig. 10.3.** Linear expansion of quartz, calcite, and the important feldspars. Plot after data from Skinner (1966)

internal stresses during moderate heating by the sun (Fig. 10.3). Table 10.1 gives the thermal expansion values for linear and volume expansion of quartz and calcite from 50 to 1000 °C.

The presence of quartz in rock is therefore most critical when heat is intense. This is in contrast with calcite, which shows a relatively small total volume increase but anomalous axial expansion and contraction. Kieslinger (1954) ascribes most fire damage to the spontaneous conversion of low to high quartz. Boyd and English's (1960) pressure-volume-temperature diagram for silica (Fig. 10.4) indicates the degree of volume expansion (decrease of density) during the process of heating by fire. The density decreases with increasing temperature but increases with increasing pressure. The lines of equal density run about parallel with the phase boundaries between the low and high quartz. The density decrease accelerates almost exponentially toward the phase boundaries to a total of about 3.7% from room temperature to the boundary line, but decreases only 1.34% across the phase boundary. Further heating and recrystallization from high quartz to tridymite and cristobalite is a slow process. The behavior of quartz is mostly

**Table 10.1.** Change in length and volume as a function of temperature

| Temperature (°C) | Change in length (%) | | | | Change in volume (%) | |
| --- | --- | --- | --- | --- | --- | --- |
| | Quartz | | Calcite | | Quartz | Calcite |
| | $\perp$ to c | // to c | $\perp$ to c | // c | | |
| 50 | 0.07 | 0.03 | | | 17 | |
| 100 | 0.14 | 0.08 | −0.042 | 0.189 | 36 | 0.105 |
| 200 | 0.30 | 0.18 | −0.096 | 0.476 | 78 | 0.285 |
| 300 | 0.49 | 0.29 | | | 1.27 | |
| 400 | 0.72 | 0.43 | −0.175 | 1.05 | 1.87 | 0.765 |
| 500 | 1.04 | 0.62 | | | 70 | |
| 570 (low-quartz) | 1.46 | 0.84 | | | 3.7 | |
| 580 (high-quartz) | 1.76 | 1.03 | | | 4.55 | |
| 600 | 1.76 | 1.02 | −0.224 | 1.843 | 4.54 | 1.395 |
| 700 | | 1.75 | 1.01 | | 4.51 | |
| 800 | | 1.73 | 0.97 | | 4.43 | |
| 900 | | 1.71 | 0.92 | | 4.34 | |
| 1000 | | 1.69 | 0.88 | | 4.26 | |

$\perp$, Perpendicular; //, parallel; c, long crystal axis.
Data compiled from Skinner (1966).

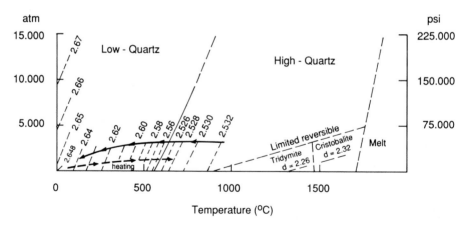

**Fig. 10.4.** Pressure-volume-temperature diagram for silica ($SiO_2$). The thick black lines mark the probable course of cooling from the liquid magma to the solid granite. *Dashed lines* mark lines of equal density. (Modified from Boyd and English 1960)

responsible for the differential expansion of the quartzes in sandstones and granites, leading to disruption, and in granites also to microcracking and a high microcrack porosity. Microcracking across and around quartz grains is described by Nur and Simmons (1970). When quartz expands again during heating by fire, it starts to exert pressure, leading to spalling. The natural porosity of granitic rocks is therefore almost twice the amount for quartz-free igneous rocks. The permissible maximum water absorption for granites is 0.4% compared with 0.2% for other igneous rocks. The high water absorption of granites enables filling of the capillaries and exposes curtain wall panels of granite to potential bowing by the expansion–contraction cycles, as has been frequently observed with the bowing of crystalline marble (Sect. 7.3). Bowing of granite panels has been observed to progress at a slower rate than for marble. To date, no published documentation is available.

Heating of quartz-rich rock in natural fires or conflagrations can cause fragmentation by differential expansion of the quartz (Figs. 10.5 & 10.6) in contrast with the adjacent rock-forming minerals (Figs. 10.2 & 10.3); loss of strength as compressive strength is plotted in Fig. 10.7 (from Ollier and Ash 1983). The presence of quartz in igneous rocks should therefore be well known and the rock checked for microcrack porosity (see also Sect. 2.9).

*Calcite*, like quartz, shows great linear expansion of almost 2% along the long or c-axis, but anomalously contracts 0.5% perpendicular to the long

**Fig. 10.5.** Fire damage by extensive spalling. *Left* Deep-reaching spalls in red quartz sandstone (Buntsandstein). Column in burnt-out building next to Mainz Cathedral (1953). *Right* Surficial spalling in limestone; column at organ choir, St. Stephen's cathedral, Vienna, Austria (1945). (Kieslinger 1954)

**Fig. 10.6.** Curved spalls from the lower course of field boulders separated by conflagration through burning paraffin candles. Quenching of the fire with cold water aggravated the contrast of temperatures near the stone surface. Fire on September 23, 1985; Lourde-type grotto, Notre Dame, Indiana

axis (Fig. 10.3) along the short or b-axis (see Table 10.1). The simultaneous expansion–contraction of crystallographically oriented calcite in marble can be disruptive, even at temperatures below 100 °C. The coefficient of expansion for calcite is therefore a function of the angle to the long axis (Fig. 10.8). Prestressed crystalline calcitic marble tends to expand more during a first cycle of heating as the locked-in stresses are released (see Fig. 7.14). Surficial calcination of carbonate rocks should start near 800 °C, and at a lower temperature for dolomites. A thin layer of burnt lime insulates the undamaged stone underneath from further excessive heating, but leaves shallow spalls and scars after the water-soluble calcium hydroxide has washed off.

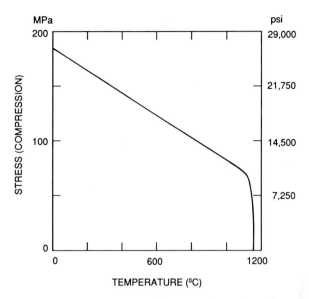

**Fig. 10.7.** Loss of rock strength with increasing exposure to elevated temperatures. (After Olliers and Ash 1983)

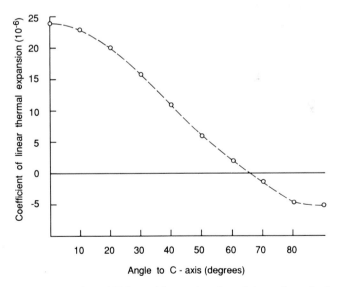

**Fig. 10.8.** Thermal expansion of Yule marble as a function of the angle to the long or c-axis. (After Rosenholtz and Smith 1949)

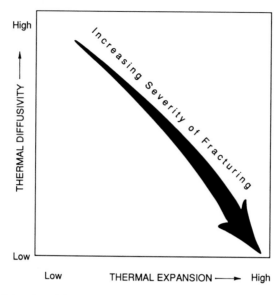

**Fig. 10.9.** Indefinite plot of the thermal diffusivity versus the thermal expansion. (After Ollier and Ash 1983)

**Table 10.2.** Effects of fire on different rock types

| Rock | Conductivity ($10^{-3}$ cal/cm) | Linear thermal expansion $\times 10^{-6}$ (%) | Critical mineral | Effect |
|------|------|------|------|------|
| Granite | 5–7 | 8 | Quartz (5–35%) | Spalling |
| Basalt | 5 | 5.4 | – | Safe |
| Limestone | 5–8 | 8 | Calcite (90–100%) | Calcining |
| Dolomite | 7–8 | | | |
| Sandstone | 5–9 | 10 | Quartz (50–100%) | Heavy spalling |
| Quartzite | 14–15 | 11 | Quartz (100%) | Spalling |
| Marble | 5–6 | 7 | Calcite (90–100%) | Calcining |
| Slate | 4–5 | 9 | Quartz (5–40%) | Spalling |

Data from Skinner (1966).

## 10.2 Conclusion

All rocks are expected to spall along the boundary of heated to unheated rock substance as the stresses become maximal along the line of greatest thermal gradient. Consequently, rocks of low thermal conductivity are more likely to fracture and spall than are rocks of high conductivity (Fig. 10.9). Table 10.2 summarizes the thermal properties and resulting damage of fire.

# References

Boyd FR, English JL (1960) The quartz-coesite transition. J Geophys Res 65(2):749–756

Freeman DC, Sadye JA, Mumpton FA (1963) The mechanism of thermal spalling in rocks. Colo School of Mines Q 58:225–252

Kieslinger A (1954) Brandeinwirkung auf Naturstein. Schweiz Arch 20:305–308

Nur A, Simmons G (1970) The origin of small cracks in igneous rocks. Int J Rock Mech Mining Sci 7(3):307–314

Ollier CD, Ash JE (1983) Fire and rock breakdown. Z Geomorphol NF 27(3):363–374

Rosenholtz JL, Smith DT (1949) Linear thermal expansion of calcite, var. Island spar, and Yule marble. Am Mineral 34:846–854

Skinner BJ (1966) Thermal expansion. In: Clark SP (ed) Handbook of physical constants. Geol Soc Am Mem 97:75–96

Underwriter's Laboratory (1964) Fire tests of building construction and materials. Standard of Safety, Underwriter's Lab Publ #263 (1959, revised 1964):22

# 11 Frost Action on Stone

Frost action on stone and concrete in moderate humid climates has long been known as a disruptive factor which deserves close attention. Frost action results from a combination of factors, such as volumetric expansion from the water to the ice phase, the degree of water saturation of the pore system, the critical pore size distribution, and the continuity of the pore system (see Sect. 6.3). The present accuracy of measurement of the pore size distribution has improved the prediction of durability. In the field, the cause of damage to stone overlaps other powerful agents, like salt action and the disruptive action of pure water below 0 °C.

## 11.1 The Process of Freezing

The pressure–temperature–volume diagram of the ice–water system has been known for over a century. Kieslinger (1930) recognized and warned of the magnitude of frost action on stone and concrete. Figure 11.1 adds contours of equal density of the ice-water phases (Winkler 1968). The updated graph offers basic theoretical information on frost danger that can be interpreted as follows. Water freezes at 0 °C and 1 atm pressure. In a confined system the pressure increases about 119 atm (1750 psi) for each °C of temperature decrease during freezing. Ice can reach a maximum pressure of 610 atm (8967 psi) at −5 °C without melting, 1130 atm (16 611 psi) at −10 °C, and 2115 atm (310 905 psi) at −22 °C, the "triple point" at which ice at a density of d = 0.948 reverts either to denser water of d = 1.086, or to the much denser ice III with d = 1.1459. The conversion of ice I to denser water thus prevents the development of greater pressures than the curve indicates. With further temperature decrease, the stress increase is only minor, with the maximum at −40 °C and 2120 atm (31 164 psi). Yet further temperature decrease results in a slight stress recovery. Ice skaters are well aware of the rapid decrease in traction on ice with decreasing temperature. The narrow edge of the skate is sufficient to melt the ice temporarily, forming a groove. Ice increases its hardness with decreasing temperature from MH = 1½ (Mohs hardness) at 0 °C to MH = 6 at −60 °C, which is the hardness of granite. The volume increase of water from 4 °C, the densest point, upward and down toward freezing appears to have some influence in

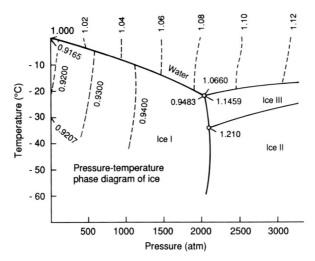

**Fig. 11.1.** The pressure–temperature–volume diagram of the ice–water system. Points of equal density are marked with *dashed lines*. (Winkler 1968)

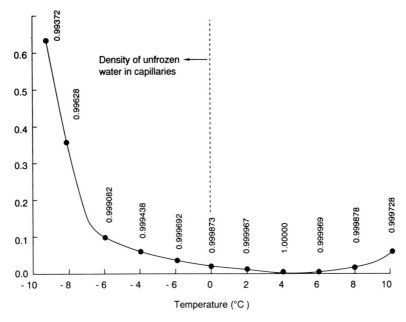

**Fig. 11.2.** Plot of density versus temperature of pure water. Note the rapidly increasing volume from +4 to −9 °C is unfrozen water. Data from Dorsey (1940)

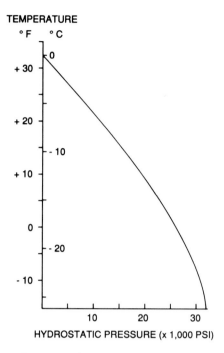

**Fig. 11.3.** Plot of the hydrostatic pressure developed in stone in advance of the freezing front, versus temperature. (Lienhart 1988)

confined capillaries. The decrease of density (or increase of volume), though minimal, causes the semiannual water turnover in small lakes. The water volume, however, increases 0.6% from freezing to −9 °C (Fig. 11.2). Bursting in narrow capillaries can thus readily occur. Unfreezable water was proven to exist at temperatures below this point by Dunn and Hudec (1966), and a more detailed account was given recently in Stockhausen et al. (1979). Lienhart (1988) has found that frost damage occurs by hydrostatic pressure caused by the inward advancing ice-front; there is a considerable increase of the hydrostatic pressure with decreasing temperature in a closed, saturated system (Fig. 11.3). It is very difficult to determine whether true crystallization of ice, or expansion of nonfreezable water, or hydrostatic pressures have caused the disintegration.

## 11.2 Susceptibility to Freezing in Capillaries

The danger to stone by frost action depends on the pore size distribution, the relative humidity (RH), the water saturation, and the possible presence of salts, especially $Na_2SO_4$ (Jerwood et al. 1990). The different kinds of

**Table 11.1.** Types of water in capillaries of varying pore size (Stockhausen et al. 1979)

| Type of water | Pore sizes (nm) | (RH) | Water freezes at (°C) |
|---|---|---|---|
| Bulk | 100 | No effect | 0 |
| Condensation | 10 | 91% | −25 |
| Oriented | 3–10 | 60–90% | −50, −20 |
| Adsorbed (2.5 mol) | 3 | NA | −160 |

water in capillaries of different pore diameter were investigated by Stockhausen et al. (1979) for concrete by use of differential thermal analysis (DTA) (Table 11.1).

Larsen and Cady's (1969) idea of water expansion from bulk water to adsorbed water was verified with Stockhausen et al.'s DTA tests (1979). Expanding water between 0 and −10 °C appears to be sufficient to disintegrate stone with a critical pore diameter (Fig. 11.2).

Specimens of stone or concrete soaked continuously before freezing are more susceptible to decay than specimens which are soaked and subsequently dried at 75 °C and 50% RH. The sensitivity of quarry-moist stone blocks is therefore not surprising (see Fig. 11.6). The curing of such blocks has been practiced since Roman times. It was well known to the English Baroque architect, Sir Christopher Wren, who cured blocks of Portland stone on the beaches of England before use. Lienhart (1981) found with systematic curing tests (number of dry–wet cycles) that the freeze–thaw durability of Berea quartz sandstone was reduced drastically from 6% weight loss to less than 0.5% after 35 freeze–thaw cycles (Fig. 11.4). It is believed to be faster in porous limestones.

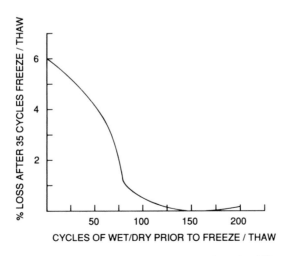

**Fig. 11.4.** Relationship between curing time and freeze – thaw durability of Berea sandstone, Ohio. (Lienhart and Stransky 1981)

**Fig. 11.5.** Dolomite in retaining wall delaminates along bedding planes, interconnected layers of clay. Frost action by the presence of adsorbed water. The wall was erected about a century ago; Milwaukee, Wisconsin

## 11.3 Freeze–Thaw Durability

The danger of frost action is especially critical on sea walls, bridges across streams, riprap, etc., where saturation above 50% of the pore spaces is possible. Frost and hygric forces readily split bedded rocks along the bedding planes at the base courses, mostly along invisible clay partings (Figs. 11.5 & 11.6). Dunn and Hudec (1966) found dolomites containing clay minerals to be more susceptible to freezing than limestones as the clay crystals tend to be more continuous in dolomites than in limestones due to "rejection textures". Such rejection textures adsorb water as unfreezable ordered water. Fine-grained rocks with a mean pore diameter of 5 µm are generally much more sensitive to frost action than are coarser grained varieties.

The outcome of laboratory testing is still somewhat uncertain. Today, we know that the pore size distribution changes toward coarser pore sizes during freezing tests, according to Fitzner (1988; Fig. 11.7). The resistance of concrete specimens to rapid freezing in air and thawing in water for 300 cycles, or until the modulus of elasticity has decreased to 60% of the sound stone, is discussed by Powers (1955) (ASTM C-290, 291-67). The freeze–thaw cycles range from +4.4 to −17.8 °C; periods range from 3 h to not more than

**Fig. 11.6.** Block of Indiana limestone resting on the floor of the quarry. Damage to the top layers of the block was probably caused by the concentration of moisture attracted to the top of the block, for maximum sensitivity to frost. Bedford, Indiana

12 h in reverse. They have been reduced to a one-cycle freezing test by Walker (1969) for rapid and fair screening. In contrast, a change can be obtained in only 12 crystallization cycles with $Na_2SO_4$ (Fig. 11.7). The contrast between the effectiveness of the freezing and the sulfate tests explains the much more rapid decay by salt action than by frost action in nature for most stones.

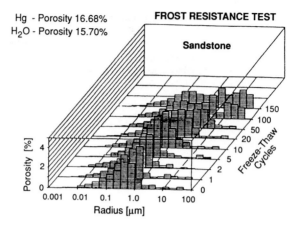

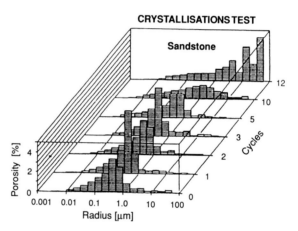

**Fig. 11.7.** Pore size distribution change during freezing tests, compared with the change during salt crystallization tests. Sander sandstone, near Mainz, Germany. (Fitzner 1988)

# References

Dorsey EN (1940) Property of ordinary water substance. Reinhold, New York, 673 pp

Dunn JR, Hudec PP (1966) Water, clay and rock soundness. Ohio J Sci 66(2):153–167

Fitzner B (1988) Porosity properties of naturally or artificially weathered sandstones. In: Ciabach J (ed) VIth Int Congr on Deterioration and conservation of stone, Torun, Sept 12–14, 1988, pp 236–245

Jerwood LC, Robinson DA, Williams RBG (1990) Experimental frost and salt weathering of chalk-II. Earth Surface Processes and Landforms 15:699–708

Kieslinger A (1930) Das Volumen des Eises. Geol Bauwesen 2:199–207

Larsen TD, Cady PD (1969) Identification of frost-susceptible particles in concrete aggregate. Natl Coop Highway Res Progr Rep 66:62

Lienhart DA, Stransky TE (1981) Evaluation of potential sources of rip-rap and armor stone–methods and considerations. Bull Assoc Engin Geol XVIII(3):323–332

Lienhart DA (1988) The geographic distribution of intensity and frequency of freeze-thaw cycles. Assoc Engin Geol Bull XXV(4):465–469

Powers TC (1955) Basic considerations pertaining to freezing and thawing tests. Am Soc Testing Materials Proc 55:1132–1155

Stockhausen N, Dorner H, Zech B, Setzer MJ (1979) Untersuchung von Gefriervorgängen in Zementstein mit Hilfe der DTA. Cement Concrete Res 9:783–794

Walker RD, Pence HJ, Hazlett WH, Ong WJ (1969) One-cycle slow-freeze test for evaluation aggregate performance in frozen concrete. Natl Coop Highway Res Progr Rep 65:21

Winkler EM (1968) Frost damage to stone and concrete: geological considerations. Eng Geol 2(5):315–323

# 12 Silicosis

Dust from mining and quarrying, as well as natural dust from dust storms and volcanic eruptions, has plagued man for many years. While the mineral components of such dust can vary to a large extent, industrial dust is usually limited to the mined product which is cut and ground in quarries and mills. Natural dusts are usually composed of clay minerals, calcite, and quartz. In contrast, industrial dusts are limited to the mined products. Dusts from coal, asbestos, calcite, and quartz have caused health problems in the past. Dust in the stone industry is usually limited to calcite and silica. While calcite dust from limestone, dolomite, and marble is fully removable from the lungs, silica dust is absorbed by the lungs and can lead to silicosis.

Silicosis is defined by the American Public Health Service (The Medical Committee on Pneumoconiosis 1937) as a dust disease due to breathing air containing silica as crystalline $SiO_2$. The symptoms are characterized by fibrotic changes in the lungs and the formation of miliary nodules in both lungs; this results in short breath, decreased chest expansion during breathing, lessened capacity for work and increased susceptibility to tuberculosis, and death. Other varieties of pneumoconiosis, like anthracosis, caused by anthracite coal dust, asbestiosis, and berylliosis, do not affect the stone industry.

The danger of silicosis increases with increasing particle surface area and decreasing size of crystalline silica as free quartz or chert, but not of amorphous opal (Fig. 12.1; Dunnom and Wagner 1981). Silicate minerals, like hornblende and feldspars, do not expose a free silica surface when silica is built into crystal lattices. Such minerals are not dangerous.

## 12.1 Dust-Producing Materials

The crystalline quartz content of commercial dusts was studied by the US Public Health Service (Table 12.1).

Some sandstones, quartzites and chert rocks can produce pure silica dust. Dusts from limestones and marble disappear from the lungs, and rather help to heal wounds inflicted by tuberculosis. Dusts from anthracite, bituminous coal, hematite, precipitator ash, and soapstone develop inert deposits in the lungs; these deposits remain in the tissues without reacting.

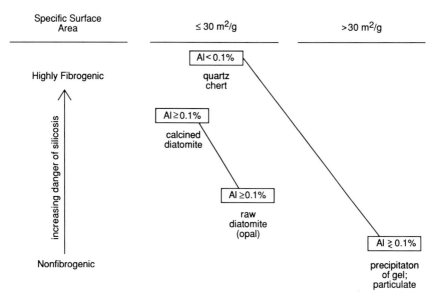

**Fig. 12.1.** Properties of silica dust as a danger for silicosis: function of grain size and specific surface area. (Dunnom and Wagner 1981)

**Table 12.1.** Dust-producing materials (After The Medical Committee 1937)

| Producer of dust | Quartz content (%) |
|---|---|
| Bituminous coal mines | 54 (sandstone in wall rock) |
| Anthracite mines | 31 (sandstone in wall rock) |
| Granite quarries | 5–35 |
| Sandstone quarries | 0–100 |
| Quartzite quarries | 100 |
| Limestone quarries | 0 to trace |
| Slate mill (Vermont red slate) | Trace |
| Slate mill (Vermont green slate) | None |

## 12.2 Dust Travel and the Physiology of Silicosis

Dust enters the lungs through the bronchial tree, the Y-junction to the main bronchus (Fig. 12.2). Lanza (1963) shows the travel of silica from the mouth to the final resting place and hyalization (glass formation) in the lungs (Fig. 12.3). It takes usually 20 years of exposure to silica dust to contract silicosis, but the time can be as short as 2.5 to 6 years (Public Health Service 1981). The result of human exposure to airborne silica dust concentrations at two

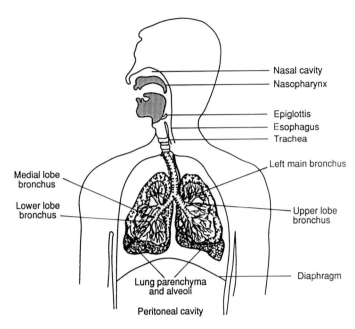

**Fig. 12.2.** Human anatomy to show travel of dust to the lungs. (Anonymous 1965)

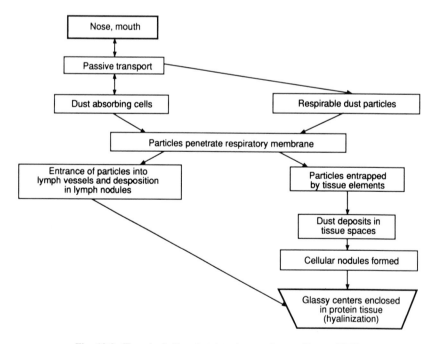

**Fig. 12.3.** Travel of silica dust into human lungs. (Lanza 1963)

silica flour mills has been recorded: 37% of those exposed contracted silicosis of the lungs in 1 to 14 years.

Free dust particles, 0.5 to 5 $\mu$m in diameter, are consumed by phagocytes in the alveoli. These phagocytes (white cells) trap and consume dust cells in the air sacs of the lungs. The nature of the dust determines the characteristics of the phagocytes, which in turn influence the symptoms of the disease. Dust rapidly scatters throughout the air passages and along the walls of the alveoli within minutes of inhalation; within the alveoli it immediately becomes ingested by phagocytes (Heppleston 1954). The phagocytes accumulate to form foci or centers and develop further to macrophages, which are major concentrations of dust-consuming cells. Some of the phagocytes and residual free dust may move back up the bronchial tree by the mechanism of alveolar clearance to become expectorated and thereby removed from the body. Klosterkoetter (1957) demonstrated on rats that physical exercise accelerates alveolar clearance. Scar tissue, stiffened by existing foci or centers, diminishes the efficiency of alveolar clearance. Persons with a history of exposure to dust are more susceptible to "dust lung" than are persons with no such disease. Further development of semielastic fibres surrounding the foci leads to emphysema. The concentration and expansion of the originally smooth air passages are shortened and narrowed during breathing. Emphysema is characterized by a shortness of breath.

A silicotic lung appears large and rigid with a pigmentation of gray to black, depending on the amount of other dusts absorbed along with silica. The three following stages of silicosis can be found:

First stage silicosis after 5.3 years;
Second stage silicosis after 6.7 years; and
Third stage silicosis with or without tuberculosis after 8 years (The Medical
    Committee 1937).

The lung appears smooth and glistening; in later stages, however, adhesions or hard nodules that are 2 to 4 mm in diameter are common. The nodules have a concentric arrangement, with reactive enclosing protein tissue around a glassy, opal-like center. The total lack of a reactive zone about a silicotic nodule speaks of a quiescent condition. Its presence indicates the progression of acute silicotic processes. Their distribution throughout the lungs may be even, or concentrated in the upper portion of the lungs. The lymph nodes are enlarged, black, and hard, adhering to the bronchi and vessels and narrowing the bronchia (bronchial stenosis). Some silica dust may free itself from the nodules and move toward the free air spaces of the lungs by alveolar clearance with the help of vibrating hairs, the cilia. New inflammations will occur by dust remaining in the alveoli or air spaces. The travel of dust in the lungs is sketched in Fig. 12.3. Severe silicotic involvement results in thickening of the air passage walls, and is called fibrosis.

About 75% of all cases of advanced silicosis are associated with tuberculosis (Gross 1959, 1963; Lanza 1963). It is still not known why tuberculosis enters the lungs and why the sputum does not disclose its presence until just before death. Tubercle infection has much decreased in recent years, and there has been a general decline of tuberculosis mortality with use of improved antibiotics.

Partial mechanical laceration by the razor-sharp edges of the quartz fragments of any size was long believed to allow the entrance of tubercles. Modern research, however, has failed to find any evidence of mechanical action by quartz dust. People with previous lung problems are likely to become victims of silicosis (Anonymous 1965). In general, the progression of silicosis is halted as soon as the exposure to silica dust stops. Lung cancer and pneumonia have not been observed as a consequence of exposure to dust.

## 12.3 Prevention of Silicosis

Careful removal of dust in quarries, mills, and building sites is the only solution for the protection of workers. Water sprinkling and spraying systems should be installed and frequently inspected. Emmerling and Seibel (1975) describe useful guidelines for spray systems in mines. New employees should be carefully screened for previous lung problems. Only approved dust masks should be used. Unfortunately, workers tend to remove face masks on hot days to make breathing easier. Inhalation of poly-2 vinyl-pyridine-N-oxide (PVNO) tends to cut the silica content in the lungs by 50% (Schlipkroeter 1969).

With care debilitating silicosis can be easily eliminated from the stone industry.

## References

Anonymous (1965) Silicosis research. Eng Mining J 166(4):98–99

Dunnom DD, Wagner P (1981) The classification of silicon dioxide powders. ASTM Standard News, November 1981, pp 10–14

Emmerling JE, Seibel RJ (1975) Dust suppression with water sprays during continuous coal mining operations. US Dep Interior Bureau of Mines Rep Invest 8064:12

Gross P (1959) Silicosis: a critique of the present concept and a proposal for modification. Am Medical Assoc Arch Indust Health 19:426–430

Gross P (1963) Pathology of the pneumoconioses. In: Lanza AJ (ed) The pneumoconioses. Grune and Stratton, New York, pp 34–58

Heppleston AG (1954) Pathogenesis of simple pneumoconiosis in coal workers. J Pathol Bacteriol 67:51–63

Klosterkoetter W (1957) Tierexperimentelle Untersuchungen über das Reinigungsvermögen der Lunge. Arch Hygiene 141:258–274

Lanza AJ (1963) The pneumoconioses. Grune and Stratton, New York, 154 pp

Public Health Service (1981) Silica Flour: silicosis. US Dep Hum Health Hum Serv, Public Health Serv, Cent for Disease Control, Current Intelligence Bull 36, June 30, 1981, 11 pp

Schlipkroeter HW (1969) PVNO – a silicosis preventative and remedy. Can Mining J 90(10):72

The Medical Committee (1937) Silicosis and allied disorders; historical and industrial importance. The Medical Committee, Air Hygiene Am, 178 pp

# 13 Stone Conservation on Buildings and Monuments

The rapid decay and disfiguring of stone monuments in urban and desert rural areas has challenged conservators to protect stone surfaces from premature or further decay. The attempt is made to halt the natural process of stone decay and possibly restore the original strength lost by chemical weathering, especially by the loss of binding cement. A general solution is not possible because the physical and chemical characteristics of an unlimited variety of stone types must be considered. The number of failures in stone preservation and attempts of restoration are greater than the number of cures.

The need for stone decay repair goes back to Roman times, for which we have evidence of replacement of decaying stone. Today, a great variety of chemicals is available for sealing paints and varnishes. Consolidants or hardeners are used to achieve the improvement of a surface weakened by chemical weathering. Attempts at sealing, however, have not been successful in the past. Undesirable stains, efflorescence, and accelerated scaling were the result. The preservation of stone falls into two very different categories, surface sealers and penetrating stone consolidants. Today, numerous chemical compounds are available to the modern conservator.

## 13.1 Sealers

Sealers develop a tight, impervious surface skin which prevents the access of moisture. Surface sealing has saved monuments from decay by eliminating the access of atmospheric humidity. Pressure tends to develop behind the stone surface by the escape of moisture. Efflorescence, crystal growth, and/or freezing can cause considerable spalling (Anderegg 1949). Flaking results when moisture is trapped behind a sealed surface, and pulled to the surface by temperature differentials. Many modern sealants bar the passage of fluids but transmit vapors, thus reducing the pressure that leads to spalling. Yellowing and blotchiness were frequently observed after use of early sealers.

The following sealants are in common use today. Linseed oil and paraffin have been in use for centuries, as have silicones, urethanes, acrylate, and even animal blood on stone and adobe. All such treatments have created more problems than cures in the past. Embrittlement, peeling, and yellowing were observed soon after application as these treatments were readily attacked by natural UV radiation.

Animal blood as paint has temporarily waterproofed masonry of adobe mud and stone masonry. Such use has a religious origin and dates back to the Phenicians and Hebrews. Instant water-soluble dried blood can be used instead of fresh blood. Winkler (1956) has described the history and techniques for the use of blood.

Silicones have proven very effective and are long lasting, while acrylates, urethane, and styrene are rapidly attacked by solar UV radiation (Clark et al. 1975). Protection against the hygric forces may require the use of waterproofing in some instances. The Egyptian granite obelisk in London may serve as an example (see Sect. 6.14; Figs. 6.38–6.40). Soon after its relocation from Egypt to London, Cleopatra's Needle was treated in 1879 with a mixture of Damar resin and wax dissolved in clear petroleum spirit, when surface scaling became evident after half a year of exposure to the humid London atmosphere. That treatment of the ancient granite monument from Egypt saved the stone from exposure to Londons's high relative humidity (RH) which would have affected the trapped salts inherited from the Egyptian desert (Burgess and Schaffer 1952). The sister obelisk in Central Park, New York City, has fared less favorably. Similar treatment was carried out too late, after the salts had hydrated and dislodged hundreds of pounds of scalings, thereby disfiguring the obelisk (Winkler 1980). Surface coating of other common stones may have to be considered for the reasons given below.

Crystalline marble: In high relative humidities curtain panels made of crystalline marble can bow when moisture is absorbed and the panels are subjected to diurnal heating and cooling cycles (Sect. 7.3). A good sealer may prevent the moisture influx provided that no moisture can enter from behind the sealer.

Limestones, dolomites: All carbonate rocks are subject to dissolution attack by rainwater, especially in areas where acid rain prevails (see Fig. 7.5). Interaction with sulfates in the atmosphere can be halted when waterproofing is used to avoid the formation of soft and more soluble gypsum. The stone surface attack is diminished if less-soluble Ca-sulfite crusts can form instead of Ca-sulfate. Replacement of calcite with fluorite or barium compounds acting as hardeners affects only the stone surface.

Sandstones: A high porosity can result in rapid water movement along a variety of pathways. Many surface sealants can do more damage by causing scaling and bursting than if no treatment were carried out at all. Sealing of any sandstone is therefore not advised.

Testing the efficiency of sealants has been the topic of several authors. Clark et al. (1975) discuss the waterproofing materials, e.g., silicones, urethanes, acrylates, and stearates, their water absorption, water vapor transmission, resistance to efflorescence, and general appearance. DeCastro

(1983) uses a microdrop (0.004 cm³) angle of contact on the stone surface to characterize the wettability as well as the exposure to UV-rich light. Laboratory tests and limited field performance are described by Heiman (1981); spalling is illustrated near the crest of a Gothic sandstone arch which was sealed with a silicone. Trapped salts are made responsible.

Sealing should be advised only in special cases where no moisture can enter from anywhere behind the treated surface.


## 13.2  Stone Consolidants

Stone solidification replaces natural grain cement lost by weathering from the stone surface inward. Replacement of crumbling stone blocks, but also treatment with waxes, linseed oil, and limewater, have been in use since ancient times. Long-lasting cures, however, were not possible. The chemical industry has developed a large variety of compounds, some inorganic but mostly organic chemicals, to strengthen stone with little or no visual changes, and has promised almost unlimited effectiveness. It is necessary to distinguish between a mere filler of open stone pores and a filler-strengthener. Inorganic and organic strengtheners are discussed by Clifton (1980), Snethlage (1984), Koblischek (1985), and others. The chemistry is discussed well by Amoroso and Fassina (1983), and by Wihr (1980) in his practical handbook for conservators and stonemasons.


### 13.2.1  Solvents

Most consolidants are solids dissolved in a solvent, the vehicle of introduction into the stone substance. Solubility, volatility, and toxicity to man are important factors for the effectiveness of a consolidant. The important solvents are discussed by Torraca (1978).

1. Aqueous: Deionized water or soft water, low in Ca, Mg, etc., to avoid efflorescing compounds that form in hard water.
2. Organic:
   a) Aliphatic hydrocarbons, e.g., mineral spirits.
   b) Aromatic hydrocarbons, e.g., benzene, toluene, xylene.
   c) Chlorinated hydrocarbons, e.g., chloroform, carbon tetrachloride, chlorothene.
   d) Alcohols, e.g., methyl alcohol, ethyl alcohol, propyl alcohol, isopropyl and butyl alcohols, glycerol, ethylene glycol.
   e) Ketones, e.g., acetone.
   f) Esters, e.g., ethyl acetate, amyl acetate.
   g) Acids, e.g., formic acid, acetic acid.

## 13.3 Inorganic Consolidants

The introduction of inorganic solutions into a crumbling stone substance, such as sandstones and crumbling limestones affected by weathering, is the natural and logical method of replacement. The following chemicals are used today with limited success: water of lime, puzzolan-cements, water-glasses, fluates, and baryte water.

### 13.3.1 Limewater Technique

Limewater is the oldest and most natural method of restoring calcitic sediments. Caroe (1987) describes the history of limewater treatment on the sculptures of the thirteenth century Wells Cathedral; these were carved in porous Doulting limestone. The soft stone soon required renovation, in which the burnt lime technique was applied, and improved through the centuries. Price (1984) experimented with limewater by applying onto the stone surface a freshly burnt lime paste, held firmly in place against the stone with sacking, to save the crumbling limestone statues above the main entrance to Wells Cathedral in England. The technique was commonly used during the Middle Ages. Price found no increase of strength, despite repeating the process. The extremely slow rate of calcification of lime mortars applied during Roman time meant that these had still not reverted completely to calcite yet (Gutschick 1988). The limited solubility of burnt lime in water [18 mg/l $Ca(OH)_2$] is another factor for the inefficiency of this natural technique (Fig. 7.5).

### 13.3.2 Fluates

The replacement of carbonates, limestone, dolostone, and crystalline marble with the colorless fluorite, $CaF_2$, is reviewed by Kessler (1883). The newly formed mineral fluorite is virtually insoluble in rainwater across the entire pH range and is harder than limestone (H = 4), in contrast to limited water-soluble calcite (H = 3). Volume-by-volume replacement from calcite to fluorite is a well-documented natural geological process which may take thousands of years, or longer. The process leaves a gray surface and was introduced about 90 years ago.

### 13.3.3 Barytization

Replacement of soluble carbonates with nearly insoluble $BaCO_3$ has been developed with urea as a catalyst, and was field tested by Bear and Lewin

(1970). Decreased solution attack and a hardened surface skin with the exchanged mineral material can be beneficial. Unfortunately, the slow reaction, the change of color, and blotchiness make this process undesirable.

### 13.3.4 Silica

The dissolution of silica in pure water depends upon the pH, the temperature of the solvent, and the degree of crystallinity (Fig. 7.4). The introduction of pure $SiO_2$ has not been successful to date. Silica, however, can be readily introduced as soluble silicates, e.g., waterglasses.

### 13.3.5 Waterglasses

The introduction of silica into crumbling quartz sandstones should replenish lost grain cement with water-soluble silicates. Commercial solutions of silica as Na or K waterglasses that leave amorphous silica were developed 150 years ago. For minimum viscosity 10 to 30% solids are used. Once the solution contacts the minerals, quartz or calcite in the stone, it solidifies to $Na_2SiO_3$ or $K_2SiO_3$ and binds loose grains within a few hours. The interaction of the waterglass with the combined $CO_2$ of the atmosphere gradually separates the alkali silicates into amorphous $SiO_2$ and $K_2CO_3$ or $Na_2CO_3$; both carbonates have very different solubilities in rainwater. Na carbonate, with limited solubility, dissolves only about $150 g/l$ and therefore tends to effloresce, while the K carbonate dissolves ten times more ($1100 g/l$), forming white efflorescence; however, it remains in solution above 40% RH, forming temporary wet areas until rain washes away the dissolved K compound. Na is rejected by most plants; it remains in circulation from the masonry into the ground and back to the masonry while the concentration increases. Residual Na or K can form compounds with salts in the masonry, resulting in combinations that form insoluble efflorescent compounds and flake. In contrast, K is absorbed as an important plant food and soon disappears from the environment. Waterglasses, however, have caused many problems in the past when they have been applied to stone. Understanding the processes will help minimize the ill effects. Table 13.1 summarizes the properties of the waterglasses.

The amorphous silica cements the grains of sand, quartz, or calcite. Extensive experiments with crumbling Carrara marble and Berea quartz sandstone, Ohio, show a substantial increase of strength by the treatment with K waterglass, more for sandstone than for marble. High ambient RH reduces the strength, much more at the beginning of the treatment than after 6 months of the progressing reaction. The initial difference in strength between oven-dried and 100% RH is ascribed to the great hygroscopicity of the K silicate. The difference diminishes rapidly after curing for a month by

**Table 13.1.** Waterglasses: pros and cons

| Waterglass | Chemical reactions | Properties | Disadvantages |
|---|---|---|---|
| $Na_2SiO_3$ | $Na_2SiO_3 + H_2O$ $= Na_2SiO_3$ (liquid) $Na_2SiO_3$ (solid) $+ CO_2$ $= SiO_2 \cdot H_2O$ (amorphous) $+ Na_2CO_3$ (150 g/l) | Very hard Hard brittle | Strong white efflorescence Hard white crusts Limited solubility |
| $K_2SiO_3$ | $K_2SiO_3 + H_2O$ $= K_2SiO_3$ (liquid) $= K_2SiO_3$ (solid) $+ CO_2$ $= SiO_2 \cdot H_2O$ (amorphous) $+ K_2CO_3$ (1100 g/l) | Similar to Na compound brittle | Occasional white efflorescence Temporary wet blotches by hygroscopic attraction |
| | Alkali-free waterglass | Proper hardening non-brittle | Slight darkening |

the gradual removal of K carbonate. Pure amorphous silica binder appears to increase the strength by gradual crystallization from the metastable amorphous silica toward the microcrystalline form. The interaction of the semiamorphous silica appears to undergo better natural binding with quartz grains than with calcite. The silica binder reduces the pore space by about one-third. Microcracking by shrinking silica gel can open up pores, leading to increased sensitivity to frost action near the surface. There is no danger of a change in color by UV radiation. Urban polluted rainwater is not likely to attack stone reinforced with silica. An alkaline solution heated by the sun, however, will attack silica gel (see Fig. 7.4). The complexity of the process of curing is reflected in the time–strength graph. High RH tends to fill the capillary system with moisture (see Sect. 6.5), reducing the strength considerably when compared with dried specimens. Sandstones tend to show patches of white or gray water-insoluble efflorescence on a colored background. Alkali-free waterglasses do not discolor the test specimens, while the advantage of a nontoxic and noncorrosive water solvent permits unlimited concentrations, and may provide possibilities for some future applications.

The geologic record suggests a similar natural existence of waterglass and the formation of feldspathic grain cement. The Brownstone sandstone of the northeastern United States has shown occasional microcrystalline feldspar cement (Hubert and Reed 1978). This cement can weather quickly to clay, losing both strength and durability. This sandstone was formed in an ancient desert valley where the presence of alkaline waters as puddles dissolved silica, alumina, and alkalis in the desert heat (see Fig. 7.4).

Natural mineral material is no doubt the most logical consolidant. The limited soluble silica or calcite still lack effective catalysts for greater solubility. Such inorganic materials are superseded by organic compounds with faster reacting and deeper penetrating solvents.

## 13.4  Organic Consolidants

Epoxides and plastics are fast becoming a very profitable material for stone consolidation methods. The proliferation of new organic consolidants with often minor variations makes impossible a discussion of all consolidants. Some silicic esters cure to inorganic hydrous amorphous silica gel when exposed to the moisture in stone pores. The most important types of organic consolidants are ethyl silicates, organic resins, and silicones. Good summaries of the organic consolidants have been compiled by Snethlage (1984), Koblischek (1985), Weber and Zinsmeister (1991), and Amoroso and Fassina (1983). Many stone conservators no longer consider a stone a "stone" when it has been treated with plastics.

### 13.4.1  Ethyl Silicates

Silicic acid ester is the consolidant used most for sandstones. The silicic acid ester reacts with moisture as follows:

$$Si(OC_2H_5)_4 + 4H_2O \rightarrow SiO_2 + 4C_2H_5OH.$$

pore water                     silica gel     ethanol

The complete reaction takes several weeks. The technique has been used successfully for several decades with good durability and no discoloring. The chemical cannot be used in dry desert areas that lack sufficient RH to assure capillary moisture for polymerization through hydrolysis. Residual moisture, on the other hand, may remain trapped behind the solidified surface zone. Shrinkage of the silica gel by gradual dehydration can cause undesirable cracks in the grain cement. The treated surface is usually pervious enough to permit the escape of moisture before flaking occurs. The pure residual silica gel resembles that of the silica deposited from waterglasses.

Several commercial products of silicic esters are presently produced under a variety of names. Zinsmeister et al. (1988) give an overview of the physical properties and testing of Ohio sandstone with silicic esters.

### 13.4.2  Silanes, Siloxenes, Siliconates

These are all characterized by forming silicon resins by hydrolysis with capillary moisture, leading from silanes possibly to siloxenes by cross linking with a silicon atom in the stone. This group of silanes requires a quartzose rock to become and remain effective.

### 13.4.3 Resins

Resins comprise a large variety of compounds and processes. The most important are epoxy resins, methacrylic resins, unsaturated polyester resins, and polyurethane resins.

Epoxy resins harden with the addition of hardeners, accelerators with active hydrogen atoms, in proportions and at temperatures that must be accurately prescribed. Though effective hardeners, epoxides lack penetration, seal out moisture, and yellow readily with exterior exposure to UV light.

Methacrylic resins are readily polymerized to polymethyl methacrylate (PMMA), generally known as plexiglass; this is resistant to weathering and to UV radiation. Accurate dosing of the ingredients is necessary to obtain favorable strength and resistance to weathering. PMMA-treated stone hardens and prevents the travel of moisture. The low viscosity of the solvent permits deep penetration and is therefore used for vacuum infiltration of small crumbling art objects. Total impregnation can only be performed on small objects in a vacuum vat and only if a statue cannot be saved in any other way. A fully impregnated natural stone, however, should no longer be called a "natural stone."

Polyester resins are not resistant to either alkalis or UV radiation. Such resins should not be used when other much better chemicals are available.

Polyurethane resins tend to form elastic polymers which do not give practical consolidants, but do provide rather favorable waterproofers. Weber and Zinsmeister (1991) and also Snethlage (1984) discuss the chemistry in more detail.

## 13.5  UV Radiation Attack

Most organic compounds tend to react with the UV part of the light spectrum in the same way as plastics. Figure 13.1 shows the solar distribution of UV radiation at summer and winter solstice for Cleveland and Chicago on the 42st parallel, Tucson and Phoenix in Arizona on the 33st parallel, and nearby Kitts Peak, 2300 m above sea level; there is a considerable summer increase in UV. The solar irradiance maxima lie in the midlatitudes. Most organic sealers and consolidants can lead to eventual crumbling after only a few years of exposure. Figure 13.2 shows a clear plastic household spoon that has become clouded and weakened by microcracks. After aging for 10 years, the numerous fine cracks have developed both by the direct destructive action of the UV and/or by crystallization from the amorphous state. Fillings of sandstones with organic consolidants may fare in a similar manner. Such microcracks increase the pore size critical to hygric, frost, and salt action after such cracks have developed in intergranular consolidant fillings.

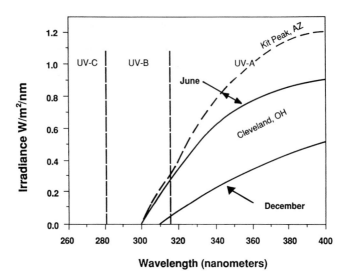

**Fig. 13.1.** Seasonal variation of UV radiation of sunlight. Midlatitudes of Cleveland, Ohio, 40 °N show great similarity with Tucson (1000 ft) and Kitts Peak, Arizona (8300 ft) (Brennan and Fedor 1987). UV-A (400 to 315 nm) causes damage to polymers and causes cataracts of the eyes; UV-B (315 to 280 nm) causes heavy polymer damage, is carcenogenic, and is absorbed by window glass

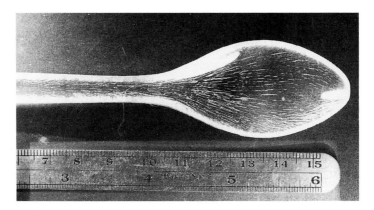

**Fig. 13.2.** Plastic household spoon of Plexiglas permeated with numerous elongated small cracks caused mostly by "aging" or crystallization of the metastable amorphous plastic. The microcracks may have also been caused by UV solar radiation. Many consolidants of this organic type may react similarly. Photo taken after 15 years of exposure

Inorganic silica is not affected by UV even if derived from organic silicic acid esters; the degree of damage varies with the compound. While some polyurethane compounds are very vulnerable, PMMA responds little and may be considered safe (Table 13.2).

**Table 13.2.** Summary of consolidants (Snethlage 1984; Koblische' '
1991)

| Chemical | Binder | Penetration | Solvent | Disc |
|---|---|---|---|---|
| *Inorganic* | | | | |
| 1. Fluoric acid | Fluates, fluorite | Limited | Water | Graying |
| $CaCO_3 + 2CaF_2 + MgSiF_6 = MgF_2 + SiO_2 + CO_2$ | | | | |
| 2. Barytization | Baryte | Limited | Water | Graying |
| $Ba(OH) + CO_2 + (urea) = BaCO_3 + H_2O$ | | | | |
| 3. Limewater | Calcite | Limited | Water | Graying |
| $Ca(OH)_2 + CO_2 = CaCO_3 + H_2O$ | | | | |
| 4. Waterglass | Silica gel | Limited | Water | |
| $Na_2SiO_3 + CO_2 = Na_2CO_3 + SiO_2$ | | | | Na compounds |
| $K_2SiO_3 + CO_2 = K_2CO_3 + SiO_2$ | | | Initial wetness | |
| *Organic* | | | | Yellowing |
| Oil, waxes | Organics | Surface | Organic solvents | Strong |
| Acrylic resins | Polymers | Limited | Organic solvents | Some | Seals, gloss |
| Epoxy resins | Polymers | Limited | Organic solvents | | Seals |
| Ethyl silicates | Silicone | Deep | Alcohol, ketone | None | |
| Silanes | Silanol (SiOH₃) | Deep | Alcohols | None | Reacts with quartz |

with = siloxenes w. $SiO_2$
Silicic acid ester

$Si(OC_2H_5) + 4H_2O = SiO_2 + alcohol$

# 13.6 Effectiveness of Consolidation

Records on the duration of effectiveness of consolidation are still scarce. The formation of spalling crusts is the major concern in the use of most chemicals. The critical problem is to achieve a uniform continuous hardness profile from the surface grading into the unweathered stone substance. The penetration of a consolidant through a case-hardened surface can be a major problem in many sandstones and pervious limestones, and it can worsen the existing situation (see Fig. 6.14).

Clarke and Ashurst (1972) studied consolidants on 24 different public buildings after exposure for 20 years to natural weathering. The results of these organic treatments were not satisfactory and it was concluded that the improved durability did not warrant the great cost. Recently, Tucci et al. (1985) infiltrated three different porous, fossiliferous limestones (Indiana limestone, Vicenza limestone, and Lecce limestone) with Acryloid B 72 (Rohm and Haas). The stones were restored to their unweathered properties and porosity. The authors have evidence from time-lapse SEM (scanning electron microscope) (REM) pictures that the solvent of the consolidants was gradually lost as a function of natural wet–dry cycles. Predictions of the durability should be based on absolute figures, e.g., per 1000 mm of rain at a pH of 4.2 and exposure to UV light.

Durability tests for consolidants should be kept simple and quantitative. Field exposures on rooftops under urban or rural conditions can approximate the expected life. Tucci et al. (1985) discuss the reduced effectiveness of acrylic and silicone resins that were separately exposed to UV and acid rain in a chamber for accelerated decay (CAD). The authors observed de-polymerization of the polymers over 3000 h of exposure to UV. Acid fog dissolved calcite crystals in limestone, leaving behind an acrylic skeletal network. Successful treatment with silicic acid ester was described by Zinsmeister et al. (1988). Quartz sandstones were carefully solidified, resulting in a pure silica gel residue. The apparent porosity, density and water absorption, modulus of rupture, compressive strength, and abrasion resistance formed the basis of the evaluation. DeWitte et al. (1984) tested 12 inorganic and organic consolidants by use of simple methods for comparison, e.g., consumption of the consolidant, impregnation depth after drying, decrease of capillary water absorption, scratch hardness, micro-sandblasting, and SEM pictures. Ethyl silicates are considered most promising, whereas the inorganic silicates are poor performers. Pure organic polymers discolor readily. The literature on the evaluation of the efficiency of consolidants and testing techniques is plentiful, yet there is no quick and practical evaluation available.

## 13.7 Recommended Testing of Consolidants

A few simple tests appear to shed some light on the durability of con-solidants. The modulus of rupture, R, has proven most valuable and involves the use of thin discs, about $34 \times 3$ mm, cored and sliced from small brick-sized stone blocks. The discs are immersed in the consolidant three times on 3 consecutive days for 3 s each time. They are tested at 2- or 3-month intervals using three methods: oven dried at 60 °C, water soaked, and exposed for 3 days to 100% RH. At least three, and preferably five, discs should be used for each test. The thin slices permit entry of the consolidants from all sides. All specimens are dipped in artificial acid rain at pH 4.2 for 1 h every month before drying or exposure to 100% RH; the progressive weight loss and the decreasing modulus of rupture are then determined (Winkler 1986). The following simple tests are summarized:

1. Water absorption on test specimens submersed for 48 h in the laboratory reflects the available total pore space, which becomes completely filled under vacuum. Water cannot enter capillaries smaller than $0.1\,\mu m$

(see Sect. 6.4), except for ordered water deposited from a high RH atmosphere.

2. Water absorption from air exposed to 100% RH for 3 days determines water absorption and reduced strength after such exposure (Fig. 2.15).

3. Permeability can be closely estimated at a stone surface with the Karsten water absorption gauge, a reliable and simple small field instrument that can be used to measure the permeability when attached to a horizontal or vertical stone surface before and after chemical treatment. The degree of absorption (mm/min or s) through the stone substance at the stone surface reflects the degree of solidification, effectiveness of a sealant with time, or a case-hardened surface. Water moves down large capillaries under the influence of gravity. Capillaries, narrowed by consolidant filling, can change the permeability and travel time of water from the tube. This water often moves upward, leaving visible wet areas around the putty at the point of attachment to the stone. Long used in Europe, Wendler and Snethlage (1989) describe the Karsten tube; Gale (1989) applies its operation to brick masonry.

4. Modulus of rupture, R, following each leaching cycle: dry, at 100% RH, and water soaked. The dry-to-wet strength ratio may serve as an index of durability (Fig. 2.16; Winkler 1986).

5. Light reflection and change of color shade compared with the US Soil Color Chart (Sect. 4.1) and a spot light meter to record the degree of darkening.

6. The porosity profile in a three-dimensional presentation (see Fig. 11.7) can show the progressive change of critical pore sizes during the process of controlled filling or leaching of consolidants, thereby recording a shift of the critical pore sizes.

Sleater (1983) extends his investigations to CAD tests, water vapor permeability, abrasion resistance, salt crystallization, salt spray, abrasion resistance, surface hardness, etc.

The tests quantify the progressive loss of the consolidant during weathering experiments, expressed especially by the pore size distribution and the loss of strength. All figures are to be compared with the untreated original stone. The literature pays much attention to comparative SEM (REM) tests which explain the kind and degree of filling within the stone fabric. Quantitative measurements, however, are possible using SEMs with image analyzers after separation of the original stone substance from the consolidants with dyes. The great variation of the stone substance, often from one layer to the next, in a natural outcrop should be considered when material is collected for testing. Successful consolidants should be applied at dilutions in which the strength and porosity do not exceed the strength of the unweathered stone substance, thereby avoiding the formation of crusts harder than the sound stone.

## 13.8 Cleaning of Stone Masonry

Cleaning buildings in and near polluted urban areas has become both an esthetic practice of building maintenance and a means of prolonging the life of the building material. Many techniques have entered the market and involve use of a multitude of modern chemicals; these have been liberally applied, usually without a proper understanding and testing of the interaction between the chemical compound and the stone substance. Riederer (1989) summarized the more important publications in a review of the existing literature. The hard sales drive of many producers of cleaners has aggravated the dilemma of the excessive use of corrosive chemicals. Such corrodents have caused much damage, e.g., accelerated scaling and discoloring. The literature is ample and includes many special reports, often sponsored by producers of chemicals and services. The techniques range from pure washing with water, high pressure steam and water-jet cleaning to the use of chemicals and abrasive methods with often very adverse results. Obviously, carbonate rocks respond differently to chemical cleaning than do silicate rocks, like quartz sandstones, or granitic rocks. Boyer (1986) and Jones (1986) discuss the state of the art of modern masonry cleaning. Several symposia deal with the problems of cleaning. The best and most practical summary and field handbook is written by Grimmer (1988) and details the labor-intensive and time-consuming techniques practiced by the cautious US National Park Service. The techniques are now accepted by most conservators and include the great problem of graffiti removal. Unbiased reports on the evaluation of cleaners have been published by several agencies, including the US National Bureau of Standards (1975), American Society for Testing Materials (ASTM) (1983), National Park Service (NPS) (1988), and Australian Mineral Development Laboratories (AMDEL) (Spry 1982; Spry and West 1985) of Australia, and articles have been devoted to special problems of cleaning. The common methods are summarized as follows:

1. *Water washing* starts with the most gentle method, involving a mild spray or gentle irrigation, enough to keep the wall wet while it softens the crusts of soot. The soot in these crusts is often intermixed with dust and gypsum. A short period of time, e.g., 3 h, is often sufficient to remove most of the deposits from the stone surface. Gentle manual brushing aids the process. Washing with increased pressure and water volume can range from mild- to high-pressure jets, with operation ranging from near 13.6 atm (200 psi) to a still safe 68 atm (1000 psi). Boyer (1986) and Jones (1984) give pressures, and water consumption at different pressures. Pressure washing can cause dissolution of carbonate rocks in areas of soft undersaturated water, forcing water inward through the capillary. High-pressure jets may also cause mechanical erosion on delicate features. Hot water can be helpful in removing grease and paint. Steam cleaning with a

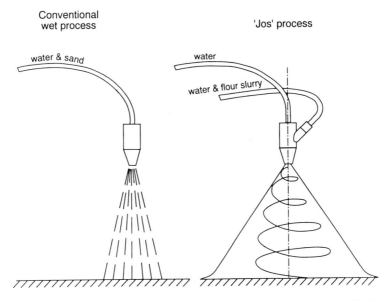

**Fig. 13.3.** Principle of JOS low-pressure rotational vortex process compared with the conventional abrasive wet-cleaning process. (Habermann 1991)

nozzle pressure of 0.5 to 5.5 bar (10 to 80 psi) is also successful when pure water pressure cannot remove the crusts.

2. *Abrasive cleaning* (dry or wet). A great number of abrasives and equipment have been developed to remove crusts and stains. Use of quartz sand blown through a nozzle at high pressure against a stone surface has become restricted by urban and environmental laws against free-flying silica dust in the atmosphere as this can cause silicosis. Milder and safer abrasives, such as glass beads, have been introduced and are in common use today; they are softer than quartz and do not endanger the lungs. Corn-husk grit and other mild abrasives are often combined with water jets. Habermann (1991) cleaned the cathedral of Regensburg with a specially designed low-pressure nozzle which mixed pure water with a slurry of glass powder in the nozzle and rotated the mixture into a turbulent, spiralling motion. This abraded the stone surface mildly, without the use of corrosive chemicals (Fig. 13.3). Mechanical abrasion tends to sculpture the stone surface, often smoothing the edges of fossil shells, and accentuating the relief between the dense, hard shells and the generally softer matrix in fossil limestones. Such a relief can be easily profiled across the surface by use of macro-stereophotogrammetry to an accuracy of 0.2 mm (Winkler 1983). Such a relief tends to trap soot and dust where it can adhere to the stone more strongly than before the treatment.

The removal of lime crusts: Structural concrete or lime mortar along joints tends to dissolve and redeposit gray crusts of lime. These new deposits are similar to dripstone formations. The strong adherence of such crusts to any natural stone surface makes their removal very difficult. Mechanical removal by scrubbing appears to be the only solution. Matero (1984) describes field tests involving a variety of commercial wet abrasives used on the Commodore Perry Memorial column on Bass Island, Ohio, with granite facing a concrete structure; multiple roof leakage supplied the water to leach the concrete and deposit the lime crusts.

3. *Laser cleaning.* Asmus et al. (1973) used laser cleaning to remove thin layers of black soot from niches and inaccessible corners of marble statues, where no other means of cleaning could reach. Long-pulse ruby laser beams burn away the black crusts in several short bursts and do not affect the white marble substrate. Soot burns at 4000 to 5000 °C. Figure 13.4 shows the mechanism of crust removal in marble. The technique has developed to a useful but time-consuming and expensive tool since its original publication in 1973.

4. *Chemical cleaning.* The chemicals fall into three categories: acidic cleaners, alkaline cleaners, and organic solvents.

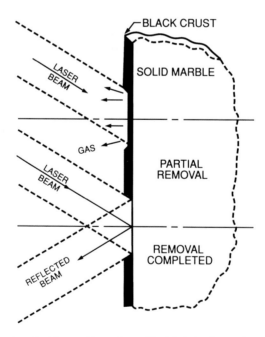

**Fig. 13.4.** Laser cleaning of white marble surfaces. The black hard crust is evaporated by short bursts of long-pulse ruby lasers, which vaporize a part of the crust. The removal is completed when the laser beam is reflected off the solid marble, after the black crust has been burned. (Adapted from Asmus et al. 1973)

Acidic cleaners include the harsh hydrofluoric acid (HF), which is generally only used with silicate rocks. Regardless of the strength of the acid, HF attacks silicate minerals, etching and corroding, and often leaving dull, nonreflecting feldspar surfaces. Eisenberg and Milton (1972) report the dulling of feldspars on a granite facade of the Old Cleveland Trust Building, Cleveland, Ohio. The HF used during the previous cleaning operation has etched the feldspars, removing the decorative sparkle of the cleavage planes. In addition, porous sandstones and limestones readily absorb and retain such acid in the pore systems, where it remains hidden and dormant. Carbonate rocks appear to be endangered the most.

Alkaline cleaners include sodium hydroxide (NaOH) and ammonium hydroxide ($NH_4OH$), the most common cleaning agents in use. Alkaline cleaners may remove surface stains, but require rinsing with water after the treatment. Careful application and thorough rinsing are very important. Iron or silica may be mobilized from inside the stone to the surface, forming either rusty blotches or gray coatings.

**Table 13.3.** Summary of cleaning techniques (After Grimmer 1988)

| Acid-sensitive masonry (e.g., limestone, marble, calcareous sandstone) | Nonacid sensitive masonry (e.g., sandstone, granites, slate) |
|---|---|
| *Dirt, Pollutant crusts* | |
|    Water + nonionic detergent | Water + detergent, acid cleaner |
| *Paint (oil, latex, acrylics)* | |
| Alkaline paint removers (ammonia or KOH, organic solvent paint) | |
| *Whitewash, Lime crusts* | |
| Acetic acid, weak HCl, wet abrasives | |
| *Stains: Iron rust* | |
|    Poultice: | Poultice: |
|       sodium citrate |    oxalic acid, phosphoric acids |
|       ammonium oxalate |    EDTA |
|    *Copper* | |
|    Poultice: | Poultice: |
|       ammonia, ammonium chloride |    ammonia, dilute HF |

*Soot, grease, tar, waxes*
  Household detergents, ammonia, bleach, mineral spirits
  Poultices: baking soda, mineral spirits, alcohol, naphtha, toluene, xylene (dry ice for
    chewing gum)

*Lichens, algae, moss, fungi*
  Dilute ammonia, bleaches, hydrogen peroxide, chloramine-T

*Graffiti*
  Organic solvents, alkaline paint remover, lacquer thinners, acetone

*Efflorescence*
  Dry brushing, washing with water, water poultices

*Bird droppings*
  Water washing (plus detergents), chelating agent (EDTA)

Organic solvents, aromatic hydrocarbons, and chlorinated hydrocarbons are often used in the removal of oil, grease, and other bituminous materials. The removal of graffiti has proven stubborn and difficult (Spry and West 1985; Godette et al. 1975). The flammability and toxicity of many organic solvents restrict their use.

## 13.9 Summary of Cleaning Techniques

Several authors have summarized pertinent material useful to the masonry cleaner, e.g., Spry (1982), Boyer (1986), and Grimmer (1988). Table 13.3 outlines methods of cleaning for different situations.

## References

Amoroso GG, Fassina V (1983) Stone decay and conservation. Materials Sci Monogr 11:453, Elsevier

Anderegg FO (1949) Testing surface water proofers. ASTM Bull TP 33:71–77

Asmus JF, Murphy CG, Munk WH (1973) Studies on the interaction of laser radiation with art artifacts. Developments in laser technology II. Proc Soc Photo-Optical Instrumentation Eng 41:19–27

Bear NS, Lewin SZ (1970) The replacement of calcite by fluorite: a kinetic study. Am Mineral 55:466–476

Boyer DW (1986) Masonry cleaning – the state of the art. In: Clifton JR (ed) Cleaning stone and masonry. ASTM Spec Tech Publ 935:25–51

Boyer DW (1987) A field and laboratory program determining the suitability of deteriorated masonries for chemical consolidation. APT Bull XIX(4):45–52

Brennan P, Fedor C (1987) Sunlight, UV and accelerated weathering. Plast Compounding 12(1):44–49

Burgess G, Schaffer RJ (1952) Cleopatra's Needle. Chem Indust October 18, 1952:1026–1029

Caroe M (1987) Conservation of figure sculptures on Wells Cathedral 1975–86. Assoc Preserv Technol XIX(4):11–15

Charola AE, Laurenzi-Tabasso M, Santamaria U (1985) The effect of water on the hydrophobic properties of an acrylic resin. In: Felix G (ed) Vth Int Congr Deterioration and conservation of stone, Lausanne, pp 739–747

Clark EJ, Campbell PG, Frohnsdorff G (1975) Waterproofing materials for masonry. Natl Bur Stand, NBS Tech Note 883:76

Clarke BL, Ashurst J (1972) Stone preservation experiments. Buildiung Res Establishment Dep Environ, 78 pp

Clifton JR (1980) Stone consolidating materials – A status report. US Dep Commer Natl Bur Stand, NBS Tech Note 1118:46

Clifton JR (1987) Preliminary performance criteria for stone treatments for the United States Capitol. US Dep Comm Natl Bur Stand NBSIR 87-3542:12

DeCastro E (1983) Studies on stone treatments. Lab Natl de Engenharia Civil, Lisboa, Memoria No 584:6

DeWitte E, Charola AE, Cherryl RP (1984) Preliminary tests on commercial stone consolidants. In: Felix G (ed) Vth Int Congr Deterioration and conservation of stone, Lausanne, Sept 25–27, 1985, pp 709–718

Eisenberg KS, Milton C (1972) Case history of a structural-granite cleaning problem. Proc 1st Int Symp Deterioration of building stones, LaRochelle, France, pp 129

Gale F (1989) Measurement of water absorption. Assoc Preserv Technol Bull XXI(3/4):8–9

Godette M, Post M, Campbell PG (1975) Graffiti-resistant coatings: methods of test and preliminary selection criteria. Natl Bur Stand US Dep Comm NBSIR 75-798:35

Grimmer AE (1984) A glossary of historic masonry deterioration and preservation treatments. Dep Interior, Natl Park Serv, Preserv Assistance Division, TH 5321 G56, 65 pp

Grimmer AE (1988) Keeping it clean. US Dep Interior, Natl Park Serv, Preserv Assistance Division, 33 pp

Gutschick KA (1988) Lime, and important construction material. ASTM Standard News, July 1988, pp 32–35

Habermann H (1991) Fassadenwäsche am Regensburger Dom. Stein 91(1):1–4

Hawkins WL (1973) Environmental deterioration of polymers. In Hawkins WL (ed) Polymer stabilization. Wiley, New York, pp 1–28

Heiman JL (1981) The preservation of Sydney sandstone by chemical impregnation. Exp Building Station, Tech Rep 469:22, Chatswood, NSW

Hubert JF, Reed AA (1978) Red-bed diagenesis in the East Berlin formation, Newark Group, Connecticut Valley. J Sediment Petrol 48:175–184

Jones LD (1986) Criteria for selection of a most appropriate cleaning method. In: Clifton JR (ed) Cleaning stone and masonry. ASTM Spec Tech Publ 935:52–70

Kessler L (1883) Sur un procédé de durcissement des pierres calcaires tendre au moyen des fluosilicates à base d'oxydes insolubles. CR Acad Sci Inst Fr 96:1317–19

Koblischek P (1985) Polymers in the renovation of buildings constructed of natural stone. In: Felix G (ed) Vth Int Congr Deterioration and conservation of stone, Lausanne, Sept 25–27, pp 749–758

Lewin SZ (1972) Recent experience with chemical techniques of stone preservation. In: Rossi-Manaresi R, Torraca G (eds) Proc of the Meet of the Joint Committee for the Conservation of Stone, The Treatment of Stone Bologna, Oct 1–3, 1971, pp 139–144

Matero FG (1984) A diagnostic study and treatment evaluation for the cleaning of Perry's Victory and International Peace Memorial. Bull Assoc Preserv Technol XVI(3, 4):39–51

Price CA (1984) The consolidation of limestone using a lime poultice and limewater. In: Petzet M (ed) Int Coll Natursteinkonservierung, Munich May 21–22, 1984, Arbeitsheft 31, pp 148–151

Riederer J (1989) Bibliography on the cleaning of stone. Rathgen-Laborberichte 1:1–5

Sleater GA (1983) Stone preservatives: Methods of laboratory testing and preliminary performance criteria. Natl Bur Stand NBS Tech Note 941:72

Snethlage R (1984) Steinkonservierung. Forschungsprogramm des Zentrallab für Denkmalpflege 1979–1983, Bayer Landesamt Denkmalpflege Munich Arbeitsheft 22:203

Spry AH (1982) Principles of cleaning masonry buildings. Aust Counc Natl Trust Victoria, Tech Bull 3(1):30

Spry AH, West DG (1985) The defense against grafiti. Aust Miner Dev Lab, Amdel Rep 1571:196

Torraca G (1978) Solubility and solvents for conservation problems. Int Centr for the Study of the Preservation and the Restoration of Cultural Property, ICCROM, Rome, 60 pp

Tucci A, Koestler RJ, Charola AE, Rossi-Manaresi R (1985) The influence of acid rain and UV radiation on the aging of acrylic and silicone resins. In: Felix G (ed) Vth Int Congr Deterioration and conservation of stone, Lausanne, Sept 25–27, 2:891–898

Weber H, Zinsmeister KJH (1991) Conservation of natural stone: Guidelines to consolidation, restoration and preservation. Expert Ehuingen, Germany, 163 pp

Wendler E, Snethlage R (1989) Der Wassereindringsprüfer nach Karsten – Anwendung und Interpretation der Messwerte. Bautenschutz Bausanierung 12:110–115

Wihr R (1980) Restaurierung von Steindenkmälern. Callwey, München, 230 pp

Winkler EM (1956) The influence of blood on clays. Soil Sci 82(3):193–200

Winkler EM (1980) Historical implications in the complexity of destructive salt weathering-cleopatra's needle, New York. Assoc Preserv Technol Bull (APT) XII(2):94–102
Winkler EM (1983) A macro stereogrammetric technique for measuring surface erosion losses on stone. In: Clifton JR (ed) Cleaning stone masonry. ASTM Spec Tech Publ 935:153–161
Winkler EM (1986) A durability index for stone. Bull Assoc Engin Geol 23:344–347
Zinsmeister KJH, Weiss NR, Gale FR (1988) Laboratory evaluation of consolidation treatment of Masillon (Ohio) sandstone. Assoc Preserv Technol Bull XX(3):35–39

# Appendix A
# Evaluation of the Soundness of Stone

First the geological history of the stone formation, e.g., depositional, magmatic, metamorphic changes, mineral composition, approx. porosity, must be explored.

## 1. Observation of Physical Damage

a) Cracks, in sandstones, marbles, and granites, caused by stress relief and uneven loading of a building (Sect. 3.2)
b) Scaling and flaking, in all rock types, by hygric action, frost or salts (Sects. 6.5, 6.6, 6.10)
c) Surface crumbling, in sandstones, some granites, and marbles. Detection of traces of efflorescent salts (Sects. 6.5–6.11)
d) Porosity, changes due to weathering, transport of grain cement, effectiveness of consolidants or sealers (Sects. 2.1, 6.4, 6.5, 6.11)
e) Ultrasound testing for quality of stone; data of weathered stone should be compared with quarry-fresh material, dry and water-soaked; ultrasound tests may replace unsightly test drilling in many instances (Sect. 2.11)
f) Moisture testing: approximate moisture content in masonry can be tested with several types of instruments in the field. The method is limited; more precision is required to determine minor quantities absorbed from high relative humidity.

## 2. Observation of Chemical Damage

a) Discoloring, yellowing, mostly by iron leached from biotite, hornblende, pyrite (yellow, ochre, green), often concentrated around or downstream of mineral grains (Sects. 4.7, 9.1).
b) Efflorescence, white to gray patches, rims or margins. Soluble halite (NaCl) from streat salting, and bitter salt or Epsom salts ($MgSO_4$): salty taste by licking test, but no salty taste of limited soluble gypsum or calcite precipitates.
c) Quick chemical semiquantitative analyses, with a few scrapings and a grain of reagent salt, $BaCl_2$, for sulfate (gypsum, bitter salt)
$$AgNO_3, \text{ for chloride (halite salt).}$$

Both salts form white precipitates in distilled water at very low concentratiöns, resemblance to milk in a clear glass test tube. Water is required for such tests, works cold or warm (Sects. 6.7, 6.8).

Avoid unsightly, destructive core drilling, unless testing for modulus of rupture or precise moisture content is needed.

### 3. Suggested Quick Testing of Stone in the Laboratory

From drill cores or small stone blocks of less than brick size:

a) Water absorption at 85 °C for 48 h; water absorption from humid atmosphere, 100% RH for 3 days in an autoclave (or a tight plastic sandwich box)
b) Modulus of rupture from discs cut from small drill cores, 35 mm or larger, tested over-dry, water-soaked, exposed to 100% RH; dry-to-wet strength ratio for prediction of durability (Sect. 2.7)
c) Porosity profile, if possible; prediction of salt attack and hygric action, frost action.

### 4. Testing of Consolidants in Stone

a) Dip thin discs 2–3 mm thick in consolidant or sealant for 10 s.
b) Leach treated discs about one a month in either distilled water or in simulated acid rain at pH 4.5, for 1 h.
c) Test specimens, at least 3 for each run, at 3-month or longer intervals, as given in No. 3, including the pore profile (Sects. 13.6, 13.7).

# Appendix B
# Properties of Some Minerals and Rocks

**Table B.1.** Properties of some rock-forming minerals

| Mineral | Mohs hardness | Specific gravity | Color | Cleavage directions | Vol. Expn., % 20°C to | | | Weathering product or solubility | Occurrence | Other properties |
|---|---|---|---|---|---|---|---|---|---|---|
| | | | | | 200 | 500 | 600 | | | |
| *Elements* | | | | | | | | | | |
| Graphite C | 1–2 | 2.2 | Black | Basal | | | | Flaky residue in soils | Metam. rocks: marble, slate, schist | |
| *Oxides* | | | | | | | | | | |
| Quartz $SiO_2$ Room temp. Above 573°C | 7 | 2.65 2.40 | White | None | 0.75 | 2.75 | 4.5 | | Granite, sandstones, quartzite | Silicosis Lack of fire resistance |
| Chert | 7 | 2.65 | White | None | Same | | | | Limestones, dolomites | Crypto-crystalline chert and |
| Opal | 5–6 | 1.9–2.2 | White | None | | | | | Sediments, weathered igneous rocks | amorphous opal expand in concrete |
| Hematite $Fe_2O_3$ | 5.5–6.5 | 5.16 | Red, black | None | 0.5 | 1.5 | | Limonite, ochre | Pigment in sediments | Hydrates slowly to limonite, rust burst |
| Magnetite $Fe_3O_4$ | 5.5 | 5.2 | Black | None | 0.5 | 1.9 | 2.3 | Limonite, ochre | Igneous rocks, metamorphic | Weathers very slowly to limonite |
| Goethite $\alpha$-FeOOH | 5.0–5.5 | 3.6–4.0 | Brown, ochre | None | | | | Weathering end product | Sediments, soils | |
| Lepidocrocite $\gamma$-FeOOH | 5 | 4.09 | Red, organe | None | | | | Weathering end product | Sediments, soils | Rare in nature |

| | | | | | | | | | | |
|---|---|---|---|---|---|---|---|---|---|---|
| Limonite $Fe_2O_3 \cdot nH_2O$ | | | Brown, ochre | None | | | | Like goethite | Sediments, soils | Limonite is goethite plus amorphous ferric hydroxide |
| *Sulfides* | | | | | | | | | | |
| Pyrite $FeS_2$ | 6.0–6.5 | 5.0–5.2 | Yellow | None | | | | Limonite | Common in all rocks | $SO_2$ ion released which may oxidize to $SO_3$. Expands during oxidation and hydration to limonite |
| Marcasite $FeS_2$ | 6.0–6.5 | 4.8–4.9 | Yellow | None | | | | Limonite | Sediments | |
| *Carbonates* | | | | | | | | | | |
| Calcite $CaCO_3$ | 3 | 2.72 | White | 3, not at right angles | 0.25 | 1.2 | 1.9 | 0.92 g/l at 20 °C at a $P_{CO_2} = 0.034\%$ | Limestones, dolomites, marbles | Solubility of all carbonates rises rapidly with increasing $P_{CO_2}$ in industrial and urban atmospheres at low temperatures |
| Dolomite $CaMg(CO_3)_2$ | 3.5–4.0 | 2.85 | White | Same as calcite | | | | Similar to calcite | Dolomites, marbles | |
| Magnesite $MgCO_3$ | 4.0–4.5 | 3.0 | White | Same as calcite | | | | Similar to calcite | Serpentines | |
| Siderite $FeCO_3$ | 4–4.5 | 3.89 | Gray | Same as calcite | | | | ca. 0.50 g/l at 20 °C at a $P_{CO_2} = 0.034\%$ | Limestones, marbles, schist | |

**Table B.1.** (Contd.)

| Mineral | Mohs hardness | Specific gravity | Color | Cleavage directions | Vol. Expn., % 20°C to 200 | 500 | 600 | Weathering product or solubility | Occurrence | Other properties |
|---|---|---|---|---|---|---|---|---|---|---|
| (Thermonatrite) $Na_2CO_3 \cdot H_2O$ | 1–1.5 | 2.26 | White | 1 | | | | Soluble | Desert floor, stone pores | Important in salt bursting |
| (Natron) $Na_2CO_3 \cdot 10H_2O$ | 1–1.5 | 1.48 | White | 2 | | | | Soluble | Desert floor, stone pores | Important in salt bursting |
| *Sulfates* | | | | | | | | | | |
| Anhydrite $CaSO_4$ | 3 | 2.95 | White | 3 | | | | 0.2–0.5 g/l | Sediments, desert floor | |
| Gypsum $CaSO_4 \cdot 2H_2O$ | 2 | 2.32 | White | 3 | | | | 1.75–2.15 g/l | Sediments, desert floor | Common in efflorescence |
| (Kieserite) $MgSO_4 \cdot H_2O$ | 3.5 | 2.57 | White | 4 | | | | Slow solution | Evaporites, desert floor, in stone | |
| (Epsomite) $MgSO_4 \cdot 7H_2O$ | 2–2.5 | 1.68 | White | 2 | | | | Slow solution | Evaporites, in stone | |
| *Silicates* Feldspars: | | | | | | | | | | |
| Orthoclase $KAlSi_3O_8$ | 6 | 2.57 | White, red | 2 | 0.15 | 0.75 | 1.1 | Kaolinite illite | Igneous, metamorphic | All feldspars are good indicators of the rock freshness |
| Albite $NaAlSi_3O_8$ | 6 | 2.60 | White | 2 | 0.45 | 1.2 | 1.45 | Kaolinite illite | | |
| Labradorite 60% $[CaAl_2Si_2O_8]$ 40% $[NaAlSi_3O_8]$ | 6 | 2.70 | Dark | 2 | 0.3 | 0.66 | 0.75 | Kaolinite illite | | |

| Mineral / Formula | | | | | | | |
|---|---|---|---|---|---|---|---|
| Muscovite $KAl_2(AlSi_3O_{10})\cdot(OH)_2$ | 2–2.5 | 2.80 | White | 1 | | Igneous, sediments, metamorphic | Very resistant to weathering |
| Biotite $K(Mg,Fe)_3(AlSi_3O_{10})(OH)_2$ | 2.5–3 | 3.1 | Black | 1 | Loss of Fe, Mg | Igneous, sediments, metamorphic | Rusty discoloration by loss of Fe |
| Kaolinite $Al_2O_3\cdot2SiO_2\cdot4H_2O$ | 2 | 2.6 | White | 1 | End product of weathering | Sediments | Absorbs water without swelling |
| Illite $KAl_2(AlSi_3O_{10})\cdot(OH)_2$ | 2.5 | 2.8 | White | 1 | End product | Sediments | "Clay-mica", similar to kaolinite |
| Montmorillonite $Al_2O_3\cdot4SiO_4\cdot nH_2O$ | 2.5 | 2–2.7 | White | 1 | End product | Sediments | Swelling clay; swelling pressure to 9000 psi |
| Glauconite $(K, Ca, Na)$, $(Al, Fe'', Fe''', Mg)_2$ $([OH]_2Si_3, AlO_{10})$ | – | 2.2–2.8 | Green | 1 | 1.57 | Sediments | Important green pigment |
| Hornblende Fe-Mg silicate | 5–6 | 3.0–3.4 | Green, black | 2 | 1.27 — Brown clay, limonite | Igneous, metamorphic | Weathers slowly; gives strength and toughness to rocks |
| Augite Fe-Mg silicate | 5.5–6 | 3.3–3.5 | Brown, green, black | 2 | 0.45 — Brown clay, limonite | Igneous, metamorphic | |

**Table B.2.** Properties of rocks in common use

| | Mohs H $H_m$ | Shore H $S_h$ | App. spec. grav. | Porosity (%) | Comp. strength ($10^3$ psi) | Modulus rupture ($10^3$ psi) | Impact toughness (in/in²) | Abrasion hardness $H_a$ | Thermal expansion ($10^{-7}$/°C) |
|---|---|---|---|---|---|---|---|---|---|
| Granites | 5.80–6.60 | 85–100 | 2.54–2.66 | 0.4–2.36 | 14–45 | 1.3–5.5 | 7–28 | 37–88 | 37–60 |
| Syenites | 5.68–6.58 | 82–99 | 2.72–2.97 | 0.9–1.9 | 27–63 | 2.3–3.2 | 6.3–14 | | (37) |
| Gabbros, diorite, diabase | 4.76–6.21 | 40–92 | 2.81–3.03 | 0.3–2.7 | 18–44 | 2–8 | 5.6–34 | | 20–30 |
| Basalt | 4–6 | 50–92 | | | 16–49 | 2–8 | 5–40 | | 22–35 |
| Limestone | 2.79–4.84 | 10–60 | 1.79–2.92 | 0.26–3.60 | 2–37 | 0.5–5.2 | 5–8.6 | 2–24 | 17–63 |
| Sandstone | 2.40–6.1 | 20–70 | | | 5–36 | 0.7–2.3 | 2–35 | 2–26 | 37–63 |
| Gneiss | 5.26–6.47 | 74–97 | 2.64–3.36 | 0.5–0.8 | 22–36 | 1.2–3.1 | 3.7–8.4 | | 13–44 |
| Quartzite | 4.2–6.6 | 55–83 | 2.75 | 0.3 | 30–91 | 1.2–4.5 | 5–30 | | 60 |
| Marble | 3.7–4.3 | 45–56 | 2.37–3.2 | 0.6–2.3 | 10–35 | 0.6–4 | 2–23 | 7–42 | 27–51 |
| Slate | | 45–58 | 2.71–2.9 | 0.1–4.3 | 20–30 | 5–16 | | 6–19 | 45–49 |
| Quartz | | | | | | | | 180 | |
| Orthoclase | | | | | | | | 53 | |
| Albite | | | | | | | | 81 | |

Compilation from Clark and Candle (1961), Windes (1949, 1950), Blair (1955, 1956).

# Appendix C
# Stone Specifications (ASTM)

ASTM (American Society for Testing and Materials) stone specifications are presented in an abridged form for granite, sandstone, limestone-marble, slate, and concrete aggregates. The specifications are suggested guidelines for stone for a given application. Almost all industrialized countries have similar standards, which may be readily obtained from the agency concerned with building codes.

## 1. Structural Granite (C615–68)

Structural granite shall include all varieties of commercial granite that are sawed, cut, split, or otherwise shaped for building purposes. Structural granite shall be classified:

*I. Engineering Grade*: bridge piers, sea and river walls, dams, and related structures, bridge superstructures, grade separations, and retaining walls, flexural members, curbstone and pavements.

*II. Architectural Grade*: Monumental structures, institutional buildings, commercial buildings, residential buildings, landscaping, parks, and other ornamental and private improvements.

Physical requirements: Structural granite shall conform to the physical requirements prescribed in Tables C.1 and C.2. Structural granite shall be sound, durable, and free from imperfections such as starts, cracks, and seams that would impair its structural integrity. Granite shall be free from minerals that will cause objectionable staining. The desired color and the permissible natural variations in color and texture shall be specified by carefully detailed description.

**Table C.1.** Physical requirements of structural granite

| Physical property | Test requirement | ASTM test |
|---|---|---|
| Absorption by weight, max., % | 0.4 | C 97 |
| Density, min., lb/ft$^3$ (kg/m$^3$) | 160 (2560) | C 97 |
| Compressive strength, min., psi (kg/mm$^2$) | 19 000 (13.4) | C 170 |
| Modulus of rupture, min., psi (kg/mm$^2$) | 1500 (1.05) | C 99 |

**Table C.2.** Physical requirements for different life expectancies (ASTM Designation C 422–58 T)

| Specific use | Life expectancy | |
|---|---|---|
| | Less than 50 years | More than 50 years |
| | Comp. strength min., psi | Comp. strength min., psi |
| *Engineering grade:* | | |
| Bridge piers, sea and river walls, dams | 25 000 | 30 000 |
| Bridge superstructures, grade separations, and retaining walls | 25 000 | 30 000 |
| Flexural members (modulus of rupture not less than 2000 psi) | 30 000 | 30 000 |
| Traffic controls, etc. | 25 000 | 30 000 |
| *Architectural grade:* | | |
| Monumental buildings | 28 000 | 30 000 |
| Institutional buildings | 26 000 | 28 000 |
| Commercial buildings | 20 000 | 26 000 |
| Residential buildings | 16 000 | 20 000 |
| Landscaping, parks, etc. | 25 000 | 30 000 |

## 2. Building Sandstone (C616–68)

The sandstone shall be free from seams, cracks, or other imperfections that would impair its structural integrity. The color desired and the permissible natural variations in color and texture shall be specified in careful detail. Sandstone-containing minerals such as pyrite and marcasite, that may upon exposure cause objectionable stain, shall be excluded. Table C.3 lists additional requirements.

**Table C.3.** Physical requirements of building sandstane

| Property | Sandstone | Quartzitic sandstone | Quartzite | ASTM test |
|---|---|---|---|---|
| Free silica content, min., % | 60 | 90 | 95 | |
| Absorption, max., % | 20 | 3 | 1 | C 97 |
| Compressive strenght, taken in weakest direction, min., psi (kg/mm$^2$) | 2000 (1.4) | 10 000 (7.0) | 20 000 (14.1) | C 170 |
| Modulus of rupture, min., psi (kg/mm$^2$) | 300 (0.2) | 1 000 (0.7) | 2 000 (1.4) | C 99 |

## 3. Dimension Limestone (C563–67)

Dimension limestone shall include stone that is sawed, cut, split, or otherwise finished or shaped, and shall specifically exclude molded, cast, or otherwise artificially aggregated units composed of fragments, and also crushed and broken stone. Dimension limestone may be classified into three categories, generally descriptive of those limestones having densities in approximate ranges:

I. (Low-density) Limestone with density ranging from 110 through 135 lb/ft³ (1.76 through 2.16 g/cm³).

II. (Medium-density) Limestone with density greater than 135 and not greater than 160 lb/ft³ (2.16 through 2.56 g/cm³).

III. (High-density) Limestone with density greater than 160 lb/ft³ (2.56 g/cm³).

Physical requirements: Dimension limestone shall be sound, durable, and free from visible defects or concentrations of materials that will cause objectionable staining or weakening under normal environments of use. Additional specifications are given in Table C.4.

**Table C.4.** Physical characteristics of dimension limestone

| Categories | Absorption, max., % | Comp. strength min., psi (kg/mm²) | Modulus rupture min., psi (kg/mm²) |
|---|---|---|---|
| I   (Low-density) | 12 | 1800 (1.25) | 400 (0.28) |
| II  (Medium-density) | 7.5 | 4000 (2.8) | 500 (0.35) |
| III (High-density) | 3 | 8000 (5.6) | 1000 (0.70) |

## 4. Exterior Marble (C503–67)

Marble is a crystalline rock composed predominantly of one or more of the following minerals: calcite, dolomite, or serpentine, and capable of taking a polish.

Physical requirements: For exterior use, marble shall be sound and free from spalls, cracks, open seams, pits, or other defects that would impair its strength, durability, or appearance. See also Table C.5.

## 5. Roofing Slate (C406–58)

For natural slate shingles as commonly used on sloping roofs and also square or rectangular tiles for flat roof coverings.

Physical requirements: Three grades are covered, based on the length of service that may be expected. Details are given in Table C.6.

**Table C.5.** Physical requirements of exterior marble

| Physical property | Requirement | ASTM test |
|---|---|---|
| Absorption, max. | 0.75% | C 97 |
| Specific gravity, min. | | |
|    calcite | 2.60 | C 97 |
|    dolomite | 2.80 | C 97 |
|    serpentine | 2.70 | C 97 |
|    travertine | 2.30 | C 97 |
| Compressive strength, min., psi (kg/mm$^2$) | 7500 (5.27) | C 170 |
| Modulus of rupture, min., psi (kg/mm$^2$) | 1000 (0.70) | C 99 |
| Abrasion resistance, min., $H_a$ | 10.0 | C 241 |

**Table C.6.** Physical requirements of roofing slate

| Designation | Service period, years | Modulus of rupture across grain, min., psi | Absorption max., % | Depth of softening mix., in. |
|---|---|---|---|---|
| Grade $S_1$ | 75 to 100 | 9000 | 0.25 | 0.002 |
| Grade $S_2$ | 40 to 75 | 9000 | 0.36 | 0.008 |
| Grade $S_3$ | 20 to 40 | 9000 | 0.45 | 0.014 |

Slates are manufactured for standard roofs, for textural roofs, for graduated roofs, and for flat roofs. The slate color shall agree closely with that of accepted samples; the following color nomenclature is used: black, blue black, gray, blue gray, purple, mottled purple and green, green, purple variegated, and weathering green (changes to buff and brown).

Imperfections: Curvature shall not exceed ⅛ in. per 12 in.

Knots and knurls are not objectionable on the top face. Slate may be rejected if the protuberances project more than ¹⁄₁₆ in. beyond the split surface.

Ribbons: Grades $S_1$ and $S_2$ shall be free from soft ribbons, and $S_3$ shall be free of soft ribbons below both nail holes.

## 6. Structural Slate (C629–68)

For general building and structural purposes.

General requirements: Slate shall be sound and free from spalls, pits, cracks or other defects. In general, slate for exterior application in ambient acidic atmosphere or in industrial areas where heavy air pollution occurs shall be free from carbonaceous ribbons. Abrasion hardness requirements pertain to slate subject to foot traffic only. Further requirements are given in Table C.7.

**Table C.7.** Physical requirements for structural slate

| Physical property | Exterior use | Interior use | ASTM |
|---|---|---|---|
| Absorption, max., % | 0.25 | 0.45 | C 121 |
| Modulus of rupture, min., psi (kg/mm²) | | | |
| across grain | 9000 (6.3) | 9000 (6.3) | C 120 |
| along grain | 7200 (5.0) | 7200 (5.0) | C 120 |
| Abrasion hardness, min, $H_a$ | 8.0 | 8.0 | C 241 |
| Acid resistance, max., in. (mm) | 0.015 (0.38) | 0.025 (0.64) | C 217 |

## 7. Concrete Aggregates (C33–67)

These specifications cover fine and coarse aggregate, other than lightweight aggregate, for use in concrete. Separate treatment is given to fine and coarse aggregate. The sieve analyses are not given here. Tables C.8 and C.9 give added limitations.

**Table C.8.** Limits for deleterious substances

| Item | Maximum % by weight of total sample | | ASTM |
|---|---|---|---|
| | Fine aggregate | Coarse aggregate | |
| Friable particles | 1.0 | 0.25 | C 142 |
| Soft particles | – | 5.0 | C 235 |
| Coal and lignite | 0.5 to 1.0 | 0.5 to 1.0 | C 123 |
| Material finer than No. 200 sieve | | | C 117 |
| subject to abrasion | 3.0 | | C 125 |
| all other concrete | 5.0 | 1.0 | |
| Chert as an impurity[a] | | | |
| severe exposure | | 1.0 | |
| mild exposure | | 5.0 | |

[a] Chert as an impurity that will disintegrate in five cycles of the soundness test, or 50 cycles of freezing and thawing in water, or that has a specific gravity, saturated-surface dry, of less than 2.35.

**Table C.9.** Physical properties of coarse aggregate for concrete

| Physical property | Gravel, crushed gravel, or crushed stone | ASTM |
|---|---|---|
| Soundness, loss in five cycles, max., % by weight | | |
| sodium sulfate | 12 | C 88 |
| magnesium sulfate | 18 | C 88 |
| Abrasion, max. loss, % by weight | 50 | C 131 |

# Appendix D
# Conversion Tables

## 1. Dimensions

| From | to | multiply by | From | to | multiply by |
|------|-----|-------------|------|-----|-------------|
| inch | centimeter | 2.54 | centimeter | feet | 0.0328 |
| inch | feet | 0.0833 | centimeter | inch | 0.3937 |
| inch | meter | 0.0254 | centimeter | meter | 0.01 |
| inch | millimeter | 25.4 | centimeter | micron | 1000 |
| inch | micron | $25.4 \times 10^3$ | centimeter | millimicron | $1 \times 10^7$ |
| feet | centimeter | 30.48 | millimeter | Ångstrom | $1 \times 10^7$ |
| feet | inch | 12.00 | millimeter | centimeter | 0.1 |
| feet | meter | 0.3048 | millimeter | feet | 0.0032 |
| micron | centimeter | 0.0001 | millimeter | inch | 0.0393 |
| micron | inch | $3.9370 \times 10^{-5}$ | millimeter | micron | 1000 |
| micron | millimeter | 1000 | | | |
| micron | millimicron | 1000 | | | |

## 2. Area

| From | to | multiply by |
|------|-----|-------------|
| square centimeter | square feet | 0.00107 |
| square centimeter | square inch | 0.1550 |
| square centimeter | square meter | 0.0001 |
| square centimeter | square millimeter | 100 |
| square millimeter | square centimeter | 0.01 |
| square millimeter | square inch | 0.00155 |
| square feet | square meter | 0.0929 |
| square inch | square centimeter | 6.4516 |
| square inch | square feet | 0.0069 |
| square inch | square millimeter | 645.16 |

## 3. Volume

| From | to | multiply by |
|------|----|-------------|
| pounds/cubic foot | gram/cubic centimeter | 0.01601 |
| pounds/cubin inch | gram/cubic centimeter | 27.6799 |
| pounds/cubic inch | gram/liter | 27.6806 |
| pound/gallon (US) | gram/cubic centimeter | 0.01198 |
| pound/gallon (US) | pound/cubic foot | 7.4805 |
| gram/cubic centimeter | gram/millimeter | 1.0000 |
| gram/cubic centimeter | pound/cubic foot | 62.4279 |
| gram/cubic centimeter | pound/cubic inch | 0.0361 |
| gram/liter | part per million (ppm) | 1000 |
| gram/milliliter | gram/cubic centimeter | 0.9999 |
| milligram/liter | part per million (ppm) | 1.000 |

## 4. Pressure

| From | to | multiply by |
|------|----|-------------|
| atmosphere | bar | 1.0132 |
| atmosphere | kilogram/centimeter square | 1.0332 |
| atmosphere | millimeter Hg/0 °C | 760 |
| atmosphere | pound/square inch | 14.6960 |
| kilogram/square centimeter | atmosphere | 0.9678 |
| kilogram/square centimeter | pound/square inch | 14.2233 |
| kilogram/square millimeter | pound/square inch | 1422.3343 |
| pound/square inch | atmosphere | 0.0680 |
| pound/square inch | gram/square centimeter | 70.3069 |
| pound/square inch | kilogram/square centimeter | 0.0703 |
| bar | atmosphere | 0.9869 |
| bar | gram/square centimeter | 1019.716 |
| bar | kilogram/square centimeter | 1.0197 |
| bar | pound/square inch | 14.5038 |

## 5. Temperature

| From | to | multiply by |
|------|-----|-------------|
| degrees Centigrade | degrees Fahrenheit | 5/9 (°F) −32 |
| degrees Fahrenheit | degrees Centigrade | §/5 (°C) +32 |
| Kelvin | degrees Centigrade | 273.15 +°C |
| degrees Centigrade | Kelvin | 273.15 −°C |

## 6. Heat Transmission

1 cal per centimeter/second/square centimeter/degree Centigrade
    = 2903 Btu per inch/hour/square foot/degree Fahrenheit
    = 241.9 Btu per foot/hour square foot/degree Fahrenheit.
1 Btu per inch/hour/square foot/degree Fahrenheit
    = 0.0003445 cal centimeter/second/square centimeter/degree Centigrade
    = 0.08333 Btu foot/hour/square foot/degree Fahrenheit.

# Appendix E
# Glossary of Geological and Technical Terms, Excluding Minerals and Architectural Terms

*Absorption.* Taking up, assimilation, or incorporation of liquids in a solid (AGI).

*Adsorption.* Adhesion of molecules of gases or molecules in solution to the surfaces of solid bodies with which they are in contact (AGI).

*Agate.* A variegated waxy quartz in which the colors are in bands, clouds or distinct groups (AGI). Term is frequently (erroneously) applied to quartz in granite or veins with bluish waxy appearance.

*Aging.* Field storage of stone block for the purpose of drying, stress relief, and case hardening.

*Aphanitic.* Pertaining to a texture of rocks in which the crystalline constituents are too small to be distinguished with the unaided eye. It includes both microcrystalline and cryptocrystalline textures (AGI).

*Aplite.* A dike rock consisting almost entirely of light-colored mineral constituents and having a characteristic fine-grained granitic texture. Aplites may range in composition from granitic to gabbroic, but when the term is used with no modifier it is generally understood to be granitic, i.e., consisting essentially of quartz and orthoclase (AGI).

*Argillite.* A rock derived either from siltstone, claystone, or shale that has undergone a somewhat higher degree of induration than is present in those rocks with an intermediate position to slate. Cleavage is approximately parallel to bedding in which it differs from slate (AGI).

*Arkose.* A rock of granular texture, formed principally by the process of mechanical aggregation. It is essentially composed of large grains of clear quartz and grains of feldspar, either lamellar or compact. These two minerals are often mixed in almost equal quantities, but oftener quartz is dominant (AGI).

*Arkosic limestone.* An impure clastic limestone containing a relatively high proportion of grains and/or crystals of feldspar, either detrital or formed in place (AGI).

*Arkosic sandstone.* A sandstone in which much feldspar is present. This may range from unassorted products of granular disintegration of fine- or medium-grained granite to a partly sorted river-laid or even marine arkosic sandstone (AGI).

*Ashlar.* Rectangular blocks having sawed, planed, or rock-faced surfaces, contrasted with cut blocks which are accurately sized and surface tooled. May be laid in courses (Stone Catalog).

*Band.* A term applied to a stratum or lamina conspicuous because it differs in color from adjacent layers; a group of layers displaying color differences is described as being banded (AGI).

*Bed.* 1. In granites and marbles a layer or sheet of the rock mass that is horizontal, commonly curved and lenticular, as developed by fractures. Sometimes also applied to the surface of parting between sheets. 2. In stratified rocks the unit layer formed by sedimentation, of variable thickness, and commonly tilted or distorted by subsequent deformation; generally develops a rock cleavage, parting, or jointing along the planes of stratification (Stone Catalog).

*Bedding plane.* In sedimentary or stratified rock, the division planes which separate the individual layers, beds, or strata (AGI).

*Blistering.* Swelling, accompanied by rupturing of a thin uniform skin, across and paralled to bedding planes in sandstones, by case hardening, and in granite by salt action and/or stress relief.

*Bluestone.* The commercial name for a dark bluish-gray feldspathic sandstone or arkose. The color is due to the presence of fine black and dark green minerals, chiefly hornblende and chlorite. The rock is extensively quarried in New York. Its toughness, due to light metamorphism, and the ease with which it may be split into thin slabs especially adapt it for use as flagstone. The term has been locally applied to other rocks, among which are dark blue slate and blue limestone (AGI).

*Breccia.* A fragmental rock whose components are angular and therefore, as distinguished from conglomerates, are not waterworn. There are friction or fault breccias, talus-breccias and eruptive (volcanic) breccias. The word is of Italian origin (AGI).

*Breccia marble.* Any marble made up of angular fragments (AGI).

*Breccia vein.* A fissure filled with fragments of rock in the interstices of which vein matter is deposited (AGI).

*Broach.* To drill or cut out material left between closely spaced drill holes (Stone Catalog).

*Brownstone.* Ferruginous sandstone in which the grains are generally coated with iron oxide. Applied almost exclusively to a dark brown sandstone derived from the Triassic of the Connecticut River Valley (AGI).

*Brushed finish.* Obtained by brushing the stone with a coarse rotary-type wire brush (Stone Catalog).

*Burst* (or Bump). See Rock burst.

*Calcite limestone.* A limestone containing not more than 5% of magnesium carbonate (ASTM).

*Calcite marble.* A crystalline variety of limestone containing not more than 5% of magnesium carbonate (ASTM).

*Calcite streaks.* Description of a white or milky-like streak occurring in stone. It is a joint plane usually wider than a glass seam and has been recemented by deposition of calcite in the crack and is structurally sound (Stone Catalog).

*Cement.* Chemically precipitated material occurring in the interstices between allogenic particles of clastic rocks. Silica, carbonates, iron oxides and hydroxides, gypsum and barite are the most common. Clay minerals and other fine clastic particles should not be considered as cement (AGI).

*Chert.* A compact, siliceous rock formed of chalcedonic or opaline silica, one or both, and of organic or precipitated origin. Chert occurs distributed through limestone, affording chert limestone. Petrographically, chert is composed of microscopic chalcedony, quartz particles, or both whose outlines range from easily resolvable to nonresolvable with the stereoscopic microscope. Particles rarely exceed 0.5 mm (AGI).

*Chroma.* The chroma notation of a color indicates the strength, the saturation, or the degree of departure of a particular hue from a neutral gray of the same value. The scales of chroma extend from 0 for a neutral gray to 10, 12, 14, depending upon the strength or saturation of the individual color.

*Clastic.* A textural term applied to rocks composed of fragmental material derived from preexisting rocks or from the dispersed consolidation products of magmas or lavas. The commonest clastic rocks are sandstones and shales, as distinct from limestone and anhydrites (AGI).

*Cleavage.* 1. The tendency of a rock to cleave or split along definite, parallel, closely spaced planes which may be highly inclined to bedding planes. It is a secondary structure, commonly confined to bedded rocks, is developed by pressure, and ordinarily is accompanied by some recrystallization of the rocks (AGI). 2. In the stone industry, cleavage is the ability of a rock to break along natural surfaces, a surface of natural parting (Stone Catalog).

*Cobblestone.* A natural rounded stone, large enough for use in paving (Stone Catalog).

*Commercial marble.* A crystalline rock composed predominantly of one or more of the following minerals: calcite, dolomite, or serpentine, and capable of taking a polish (ASTM).

*Conglomerate.* A rock made up of worn and rounded pebbles of various sizes cemented as in sandstone. It includes varieties locally known as "pudding-stone" (ASTM).

*Crossbedding.* An original lamination oblique to the main stratification. Lamination, in sedimentary rocks, confined to single beds and inclined to the general stratification (AGI).

*Crowfeet.* (Stylolite) Description of a dark gray to black zigzag marking occurring in stone. Usually structurally sound (Stone Catalog).

*Crystalline marble.* A limestone, either calcitic or dolomitic, composed of interlocking crystalline grains of the constituent minerals, and of phaneritic texture. Commonly used synonymously with "marble", and thus representing a recrystallized limestone. Improperly applied to limestones that display some obviously crystalline grains in a fine-grained mass but which are not of interlocking texture and do not compose the entire mass (Stone Catalog).

*Cut stone.* This includes all stone cut or machined to given sizes, dimension or shape, and produced in accordance with working or shop drawings which have been developed from the architect's structural drawings (Stone Catalog).

*Cutting stock.* A term used to describe slabs of varying size, finish, and thickness, and which are used in fabricating treads, risers, copings, borders, sills, stools, hearths, mantels and other special-purpose stones (Stone Catalog).

*Delamination.* Splitting of stone into laminae from the surface inward. Common in sandstones and limestones in the presence of clay layers.

*Dendrites.* A branching figure resembling a shrub or tree, produced on or in a mineral or rock by the crystallization of a foreign mineral, usually an oxide of manganese, as in moss agate; also the mineral or rock so marked (AGI).

*Diabase.* A rock of basaltic composition, consisting essentially of labradorite and pyroxene, and characterized by ophitic texture (discrete crystals or grains of pyroxene fill the interstices between lath-shaped feldspar crystals (AGI).

*Dimension stone.* Stone precut and shaped to dimensions of specified sizes (Stone Catalog).

*Diorite.* An igneous rock composed essentially of sodic plagioclase and hornblende, biotite, or pyroxene. Small amounts of quartz and orthoclase may be present (AGI).

*Dissolution.* The process of dissolving (AGI).

*Dolomite.* A limestone containing in excess of 40% magnesium carbonate as the dolomite molecule (ASTM).

*Dolomitic limestone.* See Magnesian limestone.

*Dressed or hand-dressed stone.* Rough chunks of stone cut by hand to create a square or rectangular shape. A stone which is sold as dressed stone generally refers to stone ready for installation (Stone Catalog).

*Dry joint.* An open or unhealed joint plane not filled with calcite and not structurally sound (Stone Catalog).

*Efflorescence.* A crystalline deposit appearing on stone surfaces, caused by soluble salts carried through or onto the stone by moisture, and which has sometimes been found to come from brick, tile, concrete blocks, mortar, concrete, or similar materials in the wall or above (Stone Catalog).

*Exposed aggregate.* The larger pieces of stone aggregate purposefully exposed for their color and texture in a cast slab (Stone Catalog).

*Fabric.* The orientation in space of the elements of which a rock is composed (AGI).

*Fault.* A fracture or fracture zone along which there has been displacement of the two sides relative to one another parallel to the fracture. The displacement may be a few inches or many miles (AGI).

*Fault fissure.* The fissure produced by a fault, even though it is afterward filled by a deposit of minerals (AGI).

*Fault rock.* The crushed rock due to the friction of the two walls of a fault rubbing against each other (AGI).

*Fault strike.* The direction of the intersection of the fault surface, or the shear zone, with a horizontal plane (AGI).

*Felsite.* An igneous rock with or without phenocrysts, in which either the whole or the groundmass consists of a cryptocrystalline aggregate of felsic minerals, quartz and potassium feldspar being those characteristically developed. When phenocrysts of quartz are present the rock is termed a quartz felsite, or, more commonly, a quartz porphyry (AGI).

*Field stone.* Loose blocks separated from ledges by natural processes and scattered through or upon the regolith (soil) cover; applied also to similar transported materials, such as glacial boulders and cobbles (Stone Catalog).

*Flagstone.* Thin slabs of stone used for flagging or paving walks, driveways, patios, etc. It is generally fine-grained sandstone, bluestone, quartzite, or slate, but thin slabs of other stones may be used (Stone Catalog).

*Flat joints.* See Sheeting.

*Flint.* A dense, fine-grained form of silica which is very tough and breaks with a conchoidal fracture and cutting edges. Of various colors: white yellow, gray, black. See also Chert (AGI).

*Foliation.* The laminated structure resulting from segregation of different minerals into layers parallel to the schistosity. Foliation is considered synonymous with slaty cleavage (AGI).

*Frame weathering.* Transport of dissolved salts from the inside of a stone block toward the surface, where mineral matter is redeposited near the surface of stone while the inside crumbles, leaving temporarily a frame-like structure.

*Freestone.* A sandstone which breaks easily in any direction without fracture or splitting (Stone Catalog).

*Gabbro.* A granular igneous rock rich in calcic plagioclase feldspar and hornblende or pyroxene. Dark in color, it is often marketed as black granite (ASTM).

*Gang sawed.* Description of the granular surface of stone resulting from gang sawing alone (Stone Catalog).

*Glass seam.* A joint plane in a rock that has been recemented by deposition of calcite or silica in the crack and is structurally sound (AGI).

*Gneiss.* A foliated crystalline rock composed essentially of silicate minerals with interlocking and visibly granular texture, and in which the foliation is due primarily to alternating layers, regular or irregular, of contrasting mineralogic composition. In general, a gneiss is charac-terized by relatively thick layers as compared with a schist. According to their mineralogic compositions gneisses may correspond to other rocks of crystalline, visibly granular, and interlocking texture, such as those under definition of commercial granite, and may then the known as granite-gneiss if strongly foliated, or gneissic granite if weakly foliated (ASTM).

*Gneissic granite.* Weakly foliated granite. See Granite.

*Grain.* 1. The particles or discrete crystals which comprise a rock or sediment. 2. A direction of splitting in rock, less pronounced than the rift and usually at a right angle to it (AGI).

*Granite.* A true granite is a visibly granular, crystalline rock of predominantly interlocking texture, composed essentially of alkaline feldspars and quartz. Feldspar is generally present in excess of quartz and accessory minerals (chiefly micas, hornblende, or more rarely pyroxene). The alkaline feldspar may be present as individual mineral species, as isomor-phous or mechanical intergrowth with each other, or as chemical intergrowth with the lime feldspar molecule, but 80% of the feldspar must be composed of the potash or soda feldspar molecules (ASTM).

*Granite-gneiss.* A gneiss visibly derived from granite by metamorphism. See Granite.

*Greenstone.* Includes rocks that have been metamorphosed or otherwise so altered that they have assumed a distinctive greenish color owing to the presence of one or more of the following minerals: chlorite, epidote, or actinolite (ASTM).

*Groundmass.* The material between the phenocrysts in porphyritic igneous rock. It includes the basis or base as well as the smaller crystals of the rock. Essentially synonymous with matrix (AGI).

*Grout.* Mortar of pouring consistency (Stone Cagalog).

*Hard rock.* Rock which requires drilling and blasting for its economical removal (AGI).

*Hardway.* In quarrying, especially in the quarrying of granite. The rift is the direction of easiest parting, the grain is a second direction of parting, and the hardway is the third and most difficult direction along which parting takes place (AGI).

*Head*. The end of a stone which has been tooled to match the face of the stone. Heads are used at outside corners, windows, door jams, or any place where the veneering will be visible from the side (Stone Catalog).

*Honed finish*. Honed is a superfine smooth finish (Stone Catalog).

*Honeycomb weathering*. Differential weathering of sandstone by minor differences in the resistance to weathering, also influenced by microclimates.

*Hue*. Chromatic colors in the Munsell system of color notations are divided into five principal classes which are given the hue names of red (R), yellow (Y), green (G), blue (B), and purple (P). The hues extend around a horizontal color sphere about a neutral or a chromatic vertical axis.

*Igneous*. Formed by solidification from a molten or partially molten state; one of the two great classes into which all rocks are divided, and contrasted with sedimentary. Rocks formed in this manner have also been called plutonic rocks, and are often divided for convenience into plutonic and volcanic rock, but there is no clear line between the two (AGI).

*Igneous rock series*. An assemblage of igneous rocks in a single district and belonging to a single period of igneous activity, characterized by a certain community of chemical, mineralogical, and occasionally also textural properties (AGI).

*Joint*. 1. In geology, a fracture or parting which interrupts abruptly the physical continuity of a rock mass (AGI). 2. The space between stone units, usually filled with mortar. Joints are classified as: flush, rake, cove, weathered, bead, stripped, "V" (Stone Catalog).

*Joint set*. A group of more or less parallel joints (AGI).

*Joint system*. Consists of two or more joint sets or any group of joints with a characteristic pattern, such as a radiating pattern, a concentric pattern, etc. (AGI). For classification see Natural deformation of rocks.

*Lamination*. The layering or bedding less than 1 cm in thickness in a sedimentary rock (AGI).

*Lava*. Fluid rock such as that which issues from a volcano or a fissure in the earth's surface; also the same material solidified by cooling (AGI).

*Liesegang banding*. Banding in color by diffusion (AGI).

*Liesegang rings*. Rings or bands resulting from rhythmic precipitation in a gel (AGI).

*Limestone*. A rock of sedimentary origin (including chemically precipitated material) composed principally of calcium carbonate or the double carbonate of calcium and magnesium (ASTM).

*Limestone marble*. Recrystallized limestones and compact, dense, relatively pure microcrystalline varieties that are capable of taking a polish are included in commercial marbles (ASTM).

*Lineation*. Linear parallelism or linear structure. The parallel orientation of structural features that are lines rather than planes. Lineation may be expressed by the parallel orientation of the following: long dimensions of minerals, striae on slickensides, streaks of minerals, cleavage-bedding intersection, intersection of two cleavages, and fold axis (AGI).

*Lithification*. The complex of processes that converts a newly deposited sediment into an indurated rock. It may occur shortly after deposition (AGI).

*Machine finish*. The generally recognized standard machine finish produced by the planers (Stone Catalog).

*Magmatic rock*. See Igneous rock.

*Magnesian (dolomitic) limestone*. A limestone containing not less than 5 or more than 40% of magnesium carbonate (ASTM).

*Magnesian (dolomitic) marble*. A crystalline variety of limestone containing not less than 5 or more than 40% of magnesium carbonate as the dolomite molecule (ASTM).

*Marble* (scientific definition). A metamorphic (recrystallized) limestone composed predominantly of crystalline grains of calcite or dolomite or both, having interlocking or mosaic texture. Compare with commercial marble (ASTM).

*Masonry*. Built-up construction, usually of a combination of materials set in mortar (Stone Catalog).

*Massive*. 1. Of homogeneous structure, without stratification, flow-banding, foliation, schistosity, and the like: often, but incorrectly, used as synonymous with igneous and eruptive. 2.

Occurring in thick beds, free from minor joints and lamination (applied to some sedimentary rocks).

*Metamorphic rock.* This group includes all those rocks which have formed in the solid state in response to pronounced changes of temperature, pressure, and chemical environment; these take place, in general, below the shells of weathering and cementation (AGI).

*Milky.* Fractured quartz and microcrystalline quartzites may have milky opalescent surfaces. Thick chunks of vein quartz from pegmatites constitute a popular feature stone for fireplaces, etc.

*Monzonite.* A granular plutonic rock containing approximately equal amounts of orthoclase and plagioclase, and thus intermediate between syenite and diorite. Quartz is usually present, but if it exceeds 2% by volume the rock is classified as quartz monzonite (AGI).

*Natural bed.* The setting of the stone on the same plane as it was formed in the ground. This generally applies to all stratified materials (Stone Catalog).

*Natural cleft.* When stones formed in layers are cleaved or separated along a natural seam the remaining surface is referred to as a natural cleft surface (Stone Catalog).

*Nicked-bit finish.* Obtained by planing the stone with a planer tool in which irregular nicks have been made in the cutting edge (Stone Catalog).

*Obsidian.* An ancient name for volcanic glass. Most obsidians are black, although red, green, and brown ones are known. They are often banded and normally have conchoidal fracture and a glassy luster. Most obsidians are rhyolitic in composition (AGI).

*Onyx.* 1. Translucent layers of calcite from cave deposits, often called Mexican onyx or onyx marble. 2. A cryptocrystalline variety of quartz, made up of different colored layers, chiefly white, yellow, black (AGI).

*Oolite.* A spherical to ellipsoidal body, 0.25–2.00 mm in diameter, which may or may not have a nucleus, and has concentric or radial structure or both. It is usually calcareous, but may be siliceous, hematic, or of other composition (AGI).

*Oolitic limestone.* A calcite-cemented calcareous stone formed of shell fragments, practically noncrystalline in character. It is found in massive deposits located almost entirely in Lawrence, Monroe, and Owen Counties, Indiana and in Alabama, Kansas, and Texas. This limestone is characteristically a freestone, without cleavage planes, possessing a remarkable uniformity of composition, texture, and structure. It possesses a high internal elasticity, adapting itself without damage to extreme temperature changes (Stone Catalog).

*Palletized.* A system of stacking stone on wooden pallets. Stone which comes palletized is easily moved and transported by modern handling equipment. Palletized stone generally arrives at the job site in better condition than unpalletized material (Stone Catalog).

*Pegmatite.* Those igneous rocks of coarse grain found usually as dikes associated with a large mass of plutonic rock of finer grain size. The absolute grain size is of lesser consequence than the relative size. Unless specified otherwise, the name usually means granite pegmatites, although pegmatites having gross compositions similar to other rock types are known. Some pegmatites contain rare minerals rich in such elements as lithium, boron, fluorine, niobium, tantalum, uranium, and the rare earths (AGI).

*pH.* The negative logarithm of the hydrogen ion activity (less correctly concentration). For example, pH 7 indicates an $H^+$ activity of $10^{-7}$ mol/l (AGI).

*Phaneritic.* A textural term applied to igneous rocks in which all the crystals of the essential minerals can be distinguished with the unaided eye. The adjective form phaneritic is currently used more frequently than the noun (AGI).

*Phenocryst.* A porphyritic crystal; one of the relatively large and ordinarily conspicuous crystals of the earliest generation in porphyritic igneous rocks. It has proved an extremely convenient term, although its etymology has been criticized (AGI).

*Plucked finish.* Obtained by rough planing the surface of stone, breaking or plucking out small particles to give rough texture (Stone Catalog).

*Plumose markings.* Feather-like markings on tension and extension joints. Also often found on split slate surfaces as networks resembling segments of spider webs, which start out from the chisel mark.

*Pneumoconiosis.* Dust diseases of the lungs. Includes silicosis.

*Polished finish.* The finest and smoothest finish available in stone. Generally only possible on hard, dense material (Stone Catalog). A mirror-like glossy surface which brings out the full color and character of the marble. A polished finish is usually not recommended for marbles intended to be used on building exteriors (MIA).

*Porphyritic.* A textural term for those igneous rocks in which larger crystals (phenocrysts) are set in a finer groundmass which may be crystalline or glassy, or both (AGI).

*Porphyry.* A term first given to an altered variety of porphyrite on account of its purple color, and afterward extended by common association to all rocks containing conspicuous phenocrysts in a fine-grained or aphanitic groundmass. The resulting texture is described as porphyritic. In its restricted usage, without qualification, the term porphyry usually implies a hypabyssal rock containing phenocrysts of alkali feldspar, though in the field it is generally allowed a wider scope, and commercially it is used for all porphyritic rocks (AGI).

*Quarry.* An open or surface working, usually for the extraction of building stone, as slate, limestone, etc. In its widest sense the term *mines* includes quarries, and has been sometimes so construed by the courts; but when the distinction is drawn, *mine* denotes underground workings and *quarry* denotes superficial workings (AGI).

*Quarry face.* The freshly split face of ashlar, squared off for the joints only, as it comes from the quarry, and used especially for massive work (AGI).

*Quartzite.* A quartz rock derived from sandstone, composed dominantly of quartz, and characterized by such thorough induration, either through cementation with silica or through recrystallization, that it is essentially homogeneous and breaks with vitreous surfaces that transect original grains and matrix or interstitial material with approximately equal ease. Such a stone possesses a very low degree of porosity and the broken surfaces are relatively smooth and vitreous as compared with the relatively high porosity and the dull, rough surfaces of sandstone (ASTM).

*Redox potential.* 1. Same as oxidation-reduction potential. 2. The oxidation-reduction potential of an environment; in other words, the voltage obtainable between an inert electrode placed in the environment and a normal hydrogen electrode, regardless of the particular substances present in the environment (AGI).

*Ribbon slate.* Parallel bands of variable thickness in slate which cut the slaty cleavage at an angle. Bands or ribbons are former bedding planes and may be lines of weakness; discoloring is also possible along the ribbons.

*Rift.* A planar property whereby granitic rocks split relatively easily in a direction other than the sheeting (AGI).
The most pronounced direction of splitting or cleavage of a stone. Rift and grain may be obscure, as in some granites, but are important in both quarrying and processing stone (Stone Catalog).

*Rip-rap.* Irregularly shaped stones used for facing bridge abutments and fills. Stones thrown together without order to form a foundation or sustaining wall (Stone Catalog).

*Rock burst.* A sudden and often violent failure of masses of rock in quarries, tunnels, and mines (AGI).

*Rock face.* This is similar to split face, except that the face of the stone is pitched to a given line and plane, producing a bold appearance rather than the comparatively straight face obtained in split face (Stone Catalog).

*Roundness.* The ratio of the average radius of curvature of the several corners or edges of a solid to the radius of curvature of the maximum inscribed sphere. Not to be confused with sphericity (AGI).

*Rubble.* A product term applied to dimension stone used for building purposes, chiefly walls and foundations, and consisting of irregularly shaped pieces, partly trimmed or squared, generally with one split or finished face, and selected and specified within a size range (Stone Catalog).

*Salt.* Halite, common salt. Sodium chloride (NaCl). In chemistry, and class of compounds formed when the acid hydrogen of an acid is partly or wholly replaced by a metal or a metallike radical; ferrous sulfate (FeSO₄) is an iron salt of sulfuric acid (AGI).

*Sand.* Detrital material of size range ¹⁄₁₆–2 mm diameter. Very coarse, 1–2 mm; coarse, ¹⁄₂–1 mm; medium, ¹⁄₄–¹⁄₂ mm; fine, ¹⁄₈–¹⁄₄ mm; very fine, ¹⁄₁₆ to ¹⁄₈ mm. (AGI).

*Sand-sawn finish.* The surface left as the stone comes from the gang saw. Moderately smooth, granular surface varying with the texture and grade of stone (Stone Catalog).

*Sandstone.* A cemented or otherwise compacted detrital sediment composed predominantly of quartz grains, the size grades of the latter being those of sand. Mineralogical varieties such as feldspathic and glauconitic sandstones are recognized, and also argillaceous, siliceous, calcareous, ferruginous, and other varieties according to the nature of the binding or cementing material (AGI). A sedimentary rock consisting usually of quartz cemented with silica, iron oxide, or calcium carbonate. Sandstone is durable, has a high crushing and tensile strength, and a wide range of colors and textures (Stone Catalog).

*Sawed face.* A finish obtained from the process used in producing building stone. Varies in texture from smooth to rough and coincident with the type of materials used in sawing – characterized as diamond sawn, sand sawn, shot sawn (Stone Catalog).

*Scale.* Thin lamina or paper-like sheets of rock, often loose, and interrupting an otherwise smooth surface on stone (Stone Catalog).

*Schist.* A foliated metamorphic rock (recrystallized) characterized by thin folia that are composed predominantly of minerals of thin platy or prismatic habit and whose long dimensions are oriented in approximately parallel positions along the planes of foliation. Because of this foliated structure schists split readily along these planes and so possess a pronounced rock cleavage. The more common schists are composed of the micas and mica-like minerals (such as chlorite) and generally contain subordinate quartz and/or feldspar of comparatively fine-grained texture; all gradations exist between schist and gneiss (coarsely foliated feldspathic rocks) (Stone Catalog).

*Scoria.* Irregular masses of lava resembling clinker or slag; may be cellular (vesicular), dark colored, and heavy (Stone Catalog).

*Serpentine.* A hydrous magnesium-silicate material of metamorphic origin, generally of very dark green color with markings of white, light green or black. One of the hardest varieties of natural building stone (Stone Catalog).

*Serpentine marble.* A green marble characterized by a prominent amount of the mineral serpentine (MIA).

*Shear.* A type of stress; a body is in shear when it is subjected to a pair of equal forces which are opposite in direction and which act along parallel planes (Stone Catalog).

*Sheeting.* In a restricted sense, the gently dipping joints that are essentially parallel to the ground surface; they are more closely spaced near the surface and become progressively further apart with depth. Especially well developed in granitic rocks (AGI).

*Shot-sawn finish.* A rough gang-saw finish produced by sawing with chilled-steel shot (Stone Catalog).

*Silicate.* A salt or ester of any of the silicic acids, real or hypothetical; a compound whose crystal lattice contains SiO₄ tetrahedra, either isolated or joined through one or more of the oxygen atoms to form groups, chains, sheets, or three-dimensional structures (AGI).

*Siliceous* (also, silicious). Of or pertaining to silica; containing silica, or partaking of its nature. Containing abundant quartz (AGI).

*Silicosis.* Disease of the human lungs. A condition of massive fibrosis of the lungs, marked by shortness of breath and caused by prolonged inhalation of silica dusts. See Pneumoconiosis.

*Slate.* A microgranular metamorphic rock derived from argillaceous sediments and characterized by excellent parallel cleavage, entirely independent of original bedding, by which cleavage the rock may be split easily into relatively thin slabs (ASTM).

*Soapstone.* A massive variety of talc with a soapy or greasy feel, used for hearths, washtubs, table tops, carved ornaments, chemical laboratories, etc.; known for its stain-proof qualities (Stone Catalog).

*Soft rock.* Can be removed by air-operated hammers, but cannot be handled economically by pick (AGI).

*Solubility.* The maximum concentration of solute, at a given temperature and pressure, which can be obtained by stirring the solute in the solvent; also, the concentration of the solute remaining when the temperature and pressure are changed to the given ones from values where the concentration of solute is higher. In other words, solubility is the equilibrium concentration of solute when undissolved solute is in contact with the solution. The most common units of solubility are grams of solute per 100 g of solvent and moles per liter of solution (AGI).

*Solubility product.* The equilibrium constant for the process of solution of a substance (generally in water). A high value indicates a more soluble material (AGI).

*Solution.* 1. The change of matter from the solid or gaseous state into the liquid state by its combination with a liquid; when unaccompanied by chemical change, it is called physical solution; otherwise, chemical solution. 2. The result of such change; a liquid combination of a liquid and a nonliquid substance (AGI).

*Sorting.* 1. In a genetic sense the term may be applied to the dynamic process by which material having some particular characteristic, such as similar size, shape, specific gravity, or hydraulic value, is selected from a larger heterogeneous mass. 2. In a descriptive sense, the term may be used to indicate the degree of similarity, in respect to some particular characteristic, of the component parts in a mass of material. 3. A measure of the spread of a distribution on either side of an average (AGI).

*Spall.* A stone fragment that has split or broken off (Stone Catalog).

*Sphericity.* The degree in which the shape of a fragment approaches the form of a sphere. Compare angularity, roundness (AGI).

*Split face* (sawed bed). Usually split face is sawed on the beds and is split either by hand or with a machine so that the surface face of the stone exhibits the natural quarry texture (Stone Catalog).

*Stone.* 1. Concrete earthy or mineral matter. A small piece of rock. Rock or rock-like material for building. Large natural masses of stone are generally called rocks; small or quarried masses are called stone; and the finer kinds, gravel or sand (AGI). 2. A precious stone; a gemstone (AGI). 3. Sometimes synonymous with rock, but more properly applied to individual blocks, masses, or fragments taken from their original formation or considered for commercial use (Stone Catalog).

*Strain.* Deformation resulting from applied force; within elastic limits strain is proportional to stress (AGI).

*Stratification.* A structure produced by deposition of sediments in beds or in layers (strata), lamina, lenses, wedges, and other essentially tabular units (AGI).

*Stress.* 1. Force that results in strain. 2. Resistance of a body to compressional, tensional, or torsional force (AGI).

*Stress relief.* Stresses dormant in deeply buried rock masses start to show phenomena of stress relief after erosion has removed overburden; such stress relief results in exfoliation, sheeting, rock bursts.

*Strip rubble.* Generally speaking, strip rubble comes from a ledge quarry. The beds of the stone, while uniformly straight, are of the natural cleft as the stone is removed from the ledge, and then split by machine to approximately 4 in. width (Stone Catalog).

*Stripping.* To remove from a quarry, or other open working, the overlying earth and disintegrated or barren surface rock (AGI).

*Strips.* Long pieces of stone – usually low-height ashlar courses where length to height ratio is maximum for the material used (Stone Catalog).

*Structure.* In petrology, one of the larger features of a rock mass, such as bedding, flow banding, jointing, cleavage, and brecciation; also, the sum total of such features; contrasted with texture (AGI).

*Stylolite.* A term applied to parts of certain limestones and also to some other rock, caused by pressure-solution; the "columns", of darker zigzag lines than the surrounding rock, run usually roughly parallel to the bedding planes, but may also cut across bedding planes at angles to 90°. Concentrations of clay and/or organic substance give the stylolites (also called crowfeet) a distinct color and pattern (modified from AGI).

*Syenite.* A plutonic igneous rock consisting principally of alkali feldspar, usually with one or more mafic minerals such as hornblende or biotite. The feldspar may be orthoclase, microcline, or perthite. A small amount of plagioclase may be present. Also of quartz if less than 5%. Quartz-free granite. Name from Syene (Aswan), where it was later renamed "Aswan red granite" (modified from AGI).

*Tectonic.* Of, pertaining to, or designating the rock structure and external forms resulting from the deformation of the earth's crust (AGI).

*Tectonic breccia.* An aggregation of angular coarse rocks formed as the result of tectonic movement. Included in this category are fault breccias, especially those associated with great overthrust sheets, and fold breccias (AGI).

*Tectonic conglomerate.* A coarse clastic rock produced by deformation of brittle, closely jointed rocks. Rotation of the joint block and granulation and crushing sometimes produce a rock that closely simulates a normal conglomerate (AGI).

*Terrazzo.* A type of concrete in which chips or pieces of stone, usually marble, are mixed with cement and are ground to a flat surface, exposing the chips, which take a high polish (Stone Catalog).

*Texture.* Geometrical aspects of the component particles of a rock, including size, shape, and arrangement (AGI).

*Travertine.* Calcium carbonate, of light color and usually concretionary and compact, deposited from solution in ground and surface waters. Extremely porous or cellular varieties are known as calcareous tufa, calcareous sinter, or spring deposit. Compact, banded varieties, capable of taking a polish, are called onyx marble. Travertine forms the stalactites and stalagmites of limestone caves, and the filling of some hot-spring conduits (AGI).

*Tufa.* Should not be confused with "tuff". See Travertine.

*Tuff.* A rock formed of compacted volcanic fragments, generally smaller than 4 mm in diameter (AGI).

*Veneer stone.* Any stone used as a decorative facing material which is not meant to be load bearing (Stone Catalog).

*Verde antique.* A dark green rock composed essentially of serpentine (hydrous magnesium silicate). Usually crisscrossed with white veinlets of magnesium and calcium carbonates. Used as an ornamental stone. In commerce often classed as a marble (AGI).

*Volcanic block.* A subangular, angular, round, or irregularly shaped mass of lava, varying in size up to several feet or yards in diameter (AGI).

*Volcanic breccia.* 1. A more or less indurated pyroclastic rock consisting chiefly of accessory and accidental angular ejecta 32 mm or more in diameter and lying in a fine tuff matrix. If the matrix is abundant, the term tuff breccia seems appropriate. 2. Rock composed mainly of angular volcanic fragments of either pyroclastic or detrital origin coarser than 2 mm in a matrix of any composition or texture, or with no matrix. 3. Rock composed of angular nonvolcanic fragments enclosed in a volcanic matrix (AGI).

*Vug.* 1. A cavity, often with a mineral lining of different composition from that of the surrounding rock. 2. A cavity in the rock, usually lined with a crystalline incrustation (AGI).

*Weathering.* The group of processes, such as the chemical action of air and rain-water and of plants and bacteria, and the mechanical action of changes of temperature, whereby rocks on exposure to the weather change in character, decay, and finally crumble into soil (AGI).

*Wedging.* Splitting of stone by driving wedges into planes of weakness (Stone Catalog).

*Wire saw.* A method of cutting stone by passing a twisted, multistrand wire over the stone, and immersing the wire in a slurry of abrasive material (Stone Catalog).

*Xenolith.* A term applied to allothigenous rock fragments that are foreign to the body of igneous rock in which they occur. An inclusion (AGI).

# Bibliography

Glossary of geology and related sciences (1960) Am Geol Inst, 2nd edn. Abbreviation: (AGI)

Glossary of historic masonry deterioration problems and preservation treatments (1984) Compiled by AE Grimmer, Dep Interior, Natl Park Serv, Preservation Assistance Division, Washington DC, 65 pp

Marble engineering handbook (1962) Appendix E, definitions. Marble Inst Am, Inc, Washington DC, AIA File No 8 B-1. Abbreviation: (MIA)

Standard definitions of terms relating to natural building stone (1958) Am Soc Testing Materials, ASTM Designation: C 119–50. Abbreviation: (ASTM)

Stone catalog (1968/1969) Building Stone Inst, glossary of terms. New York. Abbreviation: (Stone Catalog)

# Subject Index

# Springer-Verlag
# and the Environment

We at Springer-Verlag firmly believe that an international science publisher has a special obligation to the environment, and our corporate policies consistently reflect this conviction.

We also expect our business partners – paper mills, printers, packaging manufacturers, etc. – to commit themselves to using environmentally friendly materials and production processes.

The paper in this book is made from low- or no-chlorine pulp and is acid free, in conformance with international standards for paper permanency.

Printing: Saladruck, Berlin
Binding: Buchbinderei Lüderitz & Bauer, Berlin